THE VIEW FROM CASCADE HEAD

THE VIEW FROM CASCADE HEAD

Lessons for the Biosphere from the Oregon Coast

BRUCE A. BYERS

Oregon State University Press Corvallis

Library of Congress Cataloging-in-Publication Data

Names: Byers, Bruce Arden, author.
Title: The view from Cascade Head : lessons for the biosphere from the Oregon coast / Bruce A. Byers.
Description: Corvallis : Oregon State University Press, [2020] | Includes bibliographical references.
Identifiers: LCCN 2020036077 (print) | LCCN 2020036078 (ebook) | ISBN 9780870710353 (trade paperback) | ISBN 9780870710506 (ebook)
Subjects: LCSH: Biosphere reserves—Oregon—Cascade Head. | Cascade Head (Or.)
Classification: LCC QH76.5.O7 B94 2020 (print) | LCC QH76.5.O7 (ebook) | DDC 333.9509795—dc23
LC record available at https://lccn.loc.gov/2020036077
LC ebook record available at https://lccn.loc.gov/2020036078

∞ This paper meets the requirements of ANSI/NISO Z39.48-1992 (Permanence of Paper).

First published in 2020 by Oregon State University Press
Printed in the United States of America

Cover photograph by Duncan Berry.
Illustrations and map by Nora Sherwood.

Oregon State University Press
121 The Valley Library
Corvallis OR 97331-4501
541-737-3166 • fax 541-737-3170
www.osupress.oregonstate.edu

Contents

Cascade Head Biosphere Reserve

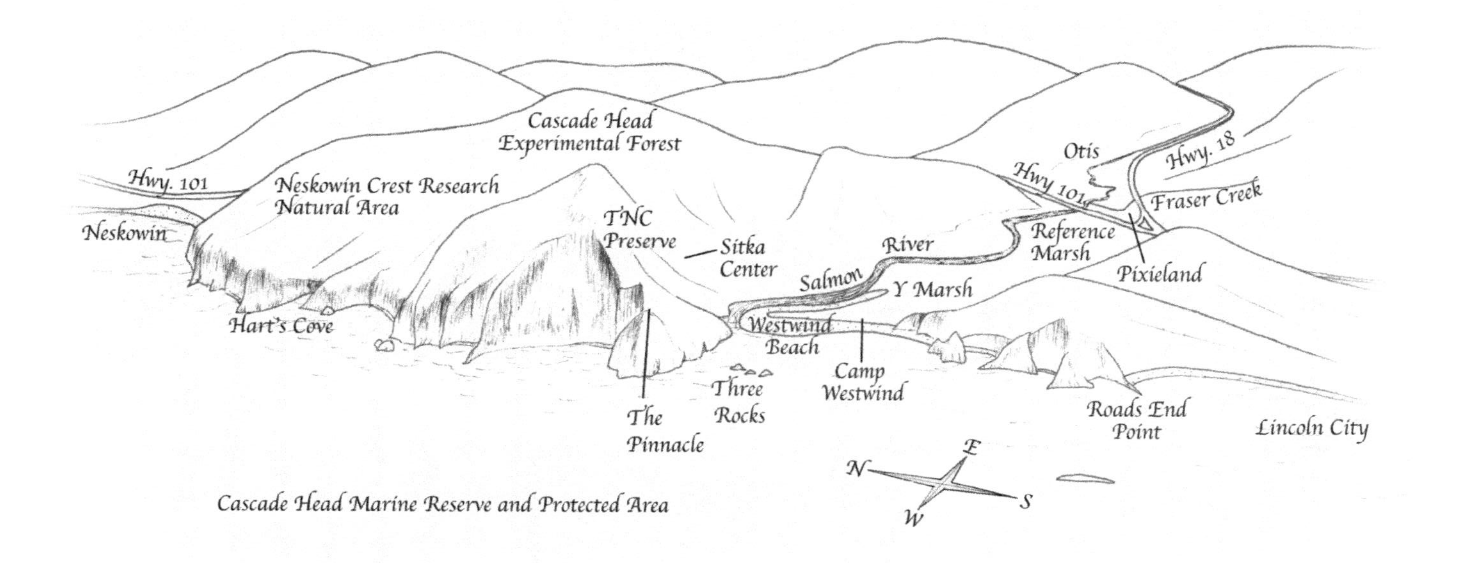

1 *Eagle's View*

> *The wild requires that we learn the terrain, nod to all the plants and animals and birds, ford the streams and cross the ridges, and tell a good story when we get back home.*
>
> —Gary Snyder, *The Practice of the Wild,* 1990

The profile of this peninsula of ancient Eocene basalt can be seen from Cape Foulweather to the south and from Cape Lookout to the north. From the east, a tattered quiltwork of clear-cuts in commercial forestry plantations scraggle down the Salmon River watershed to where forests of Sitka spruce and western hemlock rise up to the Neskowin Crest. On its western edge, the headland plunges into the Pacific over sheer cliffs cut by the relentless old ocean, and small creeks cascade from its fog-drinking forests into inaccessible coves where sea lions bark and bawl. Thus, this wild stretch of coast came to be called Cascade Head.

On its steep southern flank, above the estuary where the Salmon River slides into the ocean, coastal grasslands rise to the ridgeline. The slopes are steep, but the promise of the view is enticing. Follow the braided trails of elk across these salt-spray meadows, and suddenly you reach their edge.

A group of Oregon "conservationists" (as the newspaper called them) looked over this edge in March 1973. The *Oregonian* of Saturday, March 10, reported that just when they stepped back from the cliff, a pair of bald eagles soared overhead. The photo in the Salem *Statesman Journal* shows Senator

Robert Packwood striding up the steep southern slope of Cascade Head, his determined eye set on the high point ahead, which locals call the Pinnacle. A straggle of people, maybe a dozen, follow him, the surf-swept Three Rocks at the mouth of the Salmon River above and behind them in the photo. The caption says, "During a visit to Oregon in March, Senator Packwood led a group of conservationists and newsmen on a fact-finding hike up to the high point of Cascade Head, north of Lincoln City." The *Oregonian* story was headlined "Packwood Hopes to Preserve the Beauty of Cascade Head," with a photo that showed him sprawled at the edge of the cliff, looking down on the Pacific Ocean below.

I was looking through the meticulous files of newspaper clippings, photos, and letters that Anne Squier, one of the conservationists on that 1973 hike, had pulled out when I visited her on her houseboat home near Portland. Anne recalled that one eagle had swept up over the precipice, very close, and Malcolm Montague, the Portland lawyer and local landowner whose connections had gotten Senator Packwood involved in trying to protect Cascade Head, said, "Look, Senator! Even our national bird agrees that this area should be protected!"

Everyone on that hike that day took this visitation as a portent. The bald eagle had been our national symbol since 1782, but its numbers had plummeted because of the effects of the insecticide dichlorodiphenyltrichloroethane—DDT—which caused its eggshells, and those of many other birds of prey, to become too thin to incubate and hatch. Eagles were a "canary in the coal mine," an indicator species whose struggle to survive showed us that something we couldn't even see was tearing at the fabric of the ecosystem. When Rachel Carson's *Silent Spring* was published in 1962, it is estimated that there were only about four hundred pairs of eagles remaining in the lower forty-eight states. Seeing the eagles at Cascade Head that day must have carried a strong message: in this place, things must not be quite so bad—how can we keep it that way?

Anne's files were a time capsule documenting the political process, led by Oregon's Senator Packwood and Representative Wendell Wyatt, that created the Cascade Head Scenic Research Area (CHSRA) in 1974, a one-of-a-kind designation within the National Forest System. That legislation laid the foundation for the designation of the Cascade Head Biosphere Reserve in

1976, among the first in the United States to become part of the United Nations Educational, Scientific and Cultural Organization's (UNESCO) Man and the Biosphere Programme.

The essays that follow tell the as-yet-untold, and fascinating, story of the Cascade Head Biosphere Reserve. They grew out of my lifelong relationship with Oregon and the Oregon Coast and were catalyzed by my wanderings in the Cascade Head–Salmon River landscape during four months in the fall of 2018 as the Howard L. McKee Ecology Resident at the Sitka Center for Art and Ecology in Otis, Oregon. I tried my best to ford all the streams and cross all the ridges, greeting all the plants and animals and birds—and people—as I went. Each essay weaves what I saw and felt together with information from ecology and history, philosophical perspectives on nature conservation, and wider cross-cultural worldviews.

The international network of biosphere reserves coordinated by the UNESCO Man and the Biosphere Programme, and the concept of the "biosphere" from which it arose, are important achievements in the history of ecology, conservation, and sustainable development. Biosphere reserves are supposed to be laboratories for understanding the human-nature relationship, and models for other places to learn from as we all struggle toward a sustainable relationship between humans and our home planet. Cascade Head is Oregon's only biosphere reserve. But if you stop an Oregonian on the street and ask them whether they know about the Cascade Head Biosphere Reserve, odds are you would draw a blank look, and a question: "The what?"

The Cascade Head Biosphere Reserve is a microcosm. It is only a tiny part of our planet's thin and fragile living skin, but the efforts of many dedicated people to defend a balance between humans and nature here are illustrative and instructive. The lessons from Cascade Head apply anywhere.

Three lessons stand out. The first is the importance of individuals whose commitment, hard work, and love of place over many decades have made Cascade Head such a rich laboratory and model. Their stories are unequivocal in showing the importance of inspired, value-based, individual action. The second lesson is that although ecologists now understand much about how nature works, ecological mysteries still abound. We don't fully understand the migratory traditions of gray whales, the causes of sea star wasting syndrome, the genetic diversity of the Oregon silverspot butterfly, the life histories of salmon, or the ecohydrology of forests. More research is needed to strengthen the scientific knowledge that underpins decisions about restoring ecosystems and maintaining their resilience in the face of the changes our species is creating in the biosphere. A third lesson is the importance of worldviews—how we think about the human-nature relationship—in shaping our individual and collective actions. At Cascade Head we can read the history of changing worldviews in the landscape, and begin to imagine how a new, ecocentric worldview could create a resilient relationship between humans and nature here, and everywhere.

"We're all in this together" could be one way to state the first principle of ecology. At Cascade Head, as everywhere, the past histories and future destinies of the nonhuman and the human members of the biotic community are irrevocably intertwined.

~

The important milestones in the evolving relationship between people and nature in the Cascade Head ecosystem can be described by a handful of words with the prefix "re": resistance, research, restoration, reconciliation, and resilience. These five "re"s are common elements of efforts to heal the human-nature relationship anywhere. The essays in this book are loosely organized around the trajectory of these five elements, which the evolving story of Cascade Head exemplifies. They come up again and again throughout this book.

Resistance to actions that would have damaged or destroyed the nature of Cascade Head was an initial, critical element in its story. First came resistance against the greedy, unsustainable logging being promoted by Oregon companies and politicians, which motivated President Theodore Roosevelt and his first chief of the US Forest Service, Gifford Pinchot, to protect the area as part of a national forest in 1907. In 1974, resistance to unregulated vacation home and tourism development motivated the creation of the Cascade Head Scenic Research Area. And, in 1976, resistance to the view that human social and economic development and the conservation of nature are opposed and contradictory led to Cascade Head being designated a UNESCO biosphere reserve.

Research at Cascade Head has led to some important and widely relevant discoveries. That research was possible only because the forces that had damaged ecosystems in many other places had been resisted here. Already within the Siuslaw National Forest, a large part of Cascade Head was set aside as an experimental forest by the US Forest Service in 1934, and part of that was further protected as the Neskowin Crest Research Natural Area in 1941—a "reference" ecosystem for learning how coastal temperate rain forests function. Examples of the curiosity of scientists and the serendipity of their research are common at Cascade Head, and the long-term ecological monitoring that has occurred provides a valuable baseline for future research, including research to understand the effects of climate change.

Restoration of natural ecosystems is another hallmark of the Cascade Head story. The Cascade Head Scenic Research Area Act of 1974 provided a legal framework and some funding for the US Forest Service to begin removing dikes and tide gates and restoring natural tidal flows to areas of the Salmon River estuary that had been converted to dairy pastures starting in the 1930s. This estuarine restoration, carried out in stages beginning in 1978, created a kind of ecological experiment through which, decades later, fish biologists could study the use of the restored salt marshes by juvenile coho and Chinook salmon. When the salt marshes were reopened to the tides, juvenile salmon of both species began to feed in them immediately and to an unexpected extent, and those fish made a significant contribution to the numbers of adult salmon returning to spawn years later. The natural life-history diversity in Salmon River salmon began to reemerge because of the restoration of the estuary. Ecological restoration and research here are linked in a positive feedback loop.

Reconciliation is a term more commonly associated with social justice—such as in the post-apartheid racial healing process in South Africa—but a lot of healing is needed between humans and the biosphere too. "Biosphere reserves are about reconciling all people with the lands and waters," Eleanor Haine-Bennett, director of the Canadian National Committee for the UNESCO Man and the Biosphere Programme, told me in a phone conversation. From Cascade Head we can begin to actually see some ecological "restorative justice"—the restored salt marshes that have allowed the reemergence of life-history diversity in juvenile salmon, for example. And beavers have come back to Fraser Creek, now restored to its old channel after it was rerouted around Pixieland, a short-lived amusement park built on filled marshland along the Salmon River in the late 1960s. From Cascade Head, we can envision how restoration of the functioning natural ecosystems of a place can lead toward reconciliation of "all people with the lands and waters."

Resilience is a final "re" word in the lexicon of Cascade Head. Our home planet is dynamic and changeable, and old ideas of ecological "stability" have given way to a more sophisticated view of the dynamic balance—the resilience—of ecosystems. Think of resilience as the kind of balance it takes to ride a wave on a surfboard, not to stand still on a rock. On a planet prone to chaos, life has so far found adaptive pathways to survival, but humans have caused and accelerated global changes that now stress ecosystems in ways that threaten our own existence. If we are to survive much longer, we must rebuild the resilience of the ecosystems we have degraded. At Cascade Head, as everywhere else in the biosphere, resistance, research, restoration, and reconciliation can lead us on a path toward a more resilient future.

~

"Where do we start to resolve the dichotomy of the civilized and the wild?" Gary Snyder asked in *The Practice of the Wild*, and then a few pages later began to answer his rhetorical question by asserting that "we must first resolve to be whole." The essays that follow add up to an argument that a holistic, ecocentric worldview is the required warp for weaving a sustainable and resilient relationship between our human species and the rest of the biosphere.

Such an ecological worldview is what I call the Eagle's View. It is not new. It is the view that the Nechesne, the indigenous inhabitants of this place, saw on their vision quests on Cascade Head; that all of the people who have fought to protect the area at least glimpsed; and that we all have the opportunity to see today because of them. Thoreau, in *Walden* and many other writings, was describing such a view. So was Aldo Leopold in *A Sand County Almanac*. So, too, were John Steinbeck and Ed Ricketts in *The Log from the Sea of Cortez*. Unfortunately, this view has not yet become mainstream, and we still have much work to do to reimagine and revitalize the human-nature relationship.

The Eagle's View has ethical implications. It poses choices and asks for changes in our own personal behavior and that of our culture and society—not an easy ask. These essays sketch some of the steps that are needed to start to make us, and the ecosystems we live in, whole again. The survival of our species is at stake, yes. But on a more immediate, personal level, the meaning, or meaningfulness, of our individual lives is also at stake. That meaning is what Thoreau, Leopold, and Steinbeck and Ricketts were searching for—all of them at critical times in history, when the human-nature relationship seemed especially frayed: Thoreau when the woods around Concord and Walden Pond were being cut down as Boston sprawled in the 1840s; Leopold in the heedless land and resource depletion of the Dust Bowl era in the 1930s; and Steinbeck and Ricketts after the collapse of the California sardine fishery in the looming shadow of World War II, in the 1940s.

The vision quest for the Eagle's View is fundamentally a quest for meaning. People try to confront the fear of death and find a sense of the meaning of life through symbolic participation in something larger and more enduring, according to Robert Lee, a sociologist at the University of Washington who studied the forest management controversies of the 1980s and 1990s. He proposed four ways they look for that sense of meaning: through a focus on the biological continuity of their families; through religious beliefs in an afterlife; through their contribution to enduring creative works such as literature, science, or art; and through a feeling of being part of nature and larger cosmic processes. Environmentalists, Lee suggested, find meaning in being part of the "natural order of the world," which for them may take the place of traditional religion.

~

Just what did the group of conservationists and newspaper reporters see on that hike in 1973, I wondered? I couldn't stop thinking about that photo from the *Oregonian* that shows Senator Packwood sprawled on his belly at the edge of Cascade Head's farthest cliff, peering over. On a pure-sun day I went looking for that exact spot, that view, a copy of the old photo in hand. I sidestepped and switchbacked down from the Pinnacle to align the angles of rocks and cliffs exactly. Aha, here!

I crawled to the same edge of crumbling rock, where tiny stonecrops clung, their rosettes of succulent leaves like little pearls, above the dizzying abyss where blue water churned to white on the black rocks below.

A minute was all I could stand of that view, so full of danger and beauty, and I inched back away from that edge and stood up into the strong east wind coming down the estuary, feeling a bit dizzy, overwhelmed, and wobbly. At that exact moment, the eagle arced over on those huge dark wide wings!

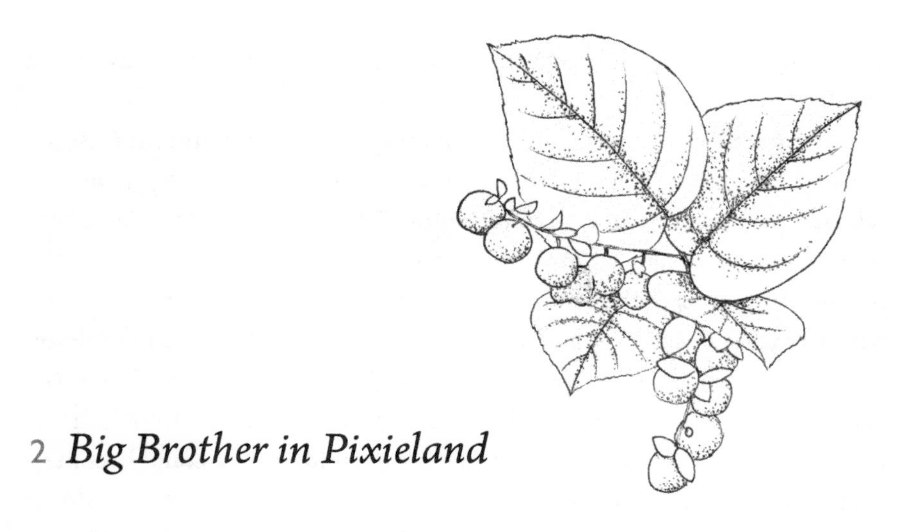

2 *Big Brother in Pixieland*

We can't pay our taxes with salal bushes!

—Jerry Parks, president of Pixieland, Inc., 1972

Newspaper clippings in Anne Squier's file show that Senator Packwood's proposal to protect Cascade Head and the Salmon River estuary wasn't uniformly popular. A clipping from the Lincoln City *News Guard* of March 16, 1972, is headlined "Landowners Differ over Cascade Head." According to the article, the audience, gathered at the Dunes Motel on March 8 to provide local feedback to the Forest Service's study,

> showed both positive and negative responses to the proposals. Jack Booth, a resident of Three Rox for over 20 years, criticized existing development in the estuary saying, "Nobody in his wildest dreams ever thought there'd be an amusement park" in the middle of the Salmon River estuary. He urged prevention of future growth of that type.

But Jerry Parks, the developer and owner of Pixieland, the amusement park that Mr. Booth was complaining about, "said he had no quarrel with environmentalists, but that 'we can't pay our taxes with salal bushes!' He opposed governmental management of the study area, arguing that it is 'such a creeping thing,' He said, further, 'we don't need Big Brother to advise and regulate.'"

The Forest Service's report on acquiring private land around Cascade Head, the *Cascade Head–Salmon River Land Use and Ownership Plan,* was released four months later in Washington, DC, by Senator Packwood and Representative Wyatt, who planned to introduce a bill to implement the recommendations of the report. The report stated that the "unique combination of natural values" found in and around Cascade Head and the Salmon River are "unparalleled on the Oregon coast," and warned that "if commercial recreation and private developments are allowed to continue at their present rate, this delicate estuary and the surrounding area could become a conglomeration of incompatible land uses," destroying the things that attract people here. The maintenance of this area in its near natural state is "an obligation to the American people," the report concluded.

Although Packwood and Wyatt introduced their bill for the Cascade Head Scenic Research Area in September of 1972, it did not pass that year. Richard Nixon was reelected president in a landslide that fall, and Democrats won control of both houses of Congress (both Packwood and Wyatt were Republicans). The bill was reintroduced in the Ninety-Third Congress on June 4, 1973, and finally—after a few national distractions (Vietnam, Watergate) that overshadowed their efforts—the Cascade Head Scenic Research Area Act, Public Law 93-535, was signed into law by President Gerald Ford on December 22, 1974.

~

I heard the roar of Frank's big wine-colored twin-cab pickup as he drove up into the courtyard just below my house on the Sitka Center campus. It was 9:28 on a chilly, quiet, gray Sunday morning—Frank was very punctual for the informal conversation we'd scheduled for about 9:30. Frank Boyden, like Anne Squier, had been part of the raggle-taggle band that accompanied Packwood to the Cascade Head cliffs on that March day in 1973, his head of youthful dark hair in about the middle of the group in the Salem *Statesman Journal* photo. I had invited him for Sunday morning coffee because I wanted to hear his recollections about that day and pick his memory about much, much more. He came in and took off his shoes, stirred sugar and a big dose of half-and-half into the promised cup of coffee, and we sat in the living room of Morley House—my comfortable abode at the Sitka Center. The Sitka Center, a small and unique nonprofit organization, was founded at the beginning of the 1970s by Frank and his wife Jane. Its quiet rural campus, sheltered by giant old Sitka spruce trees for which it is named, is situated near the mouth of the Salmon River on the southern flank of Cascade Head.

After quick updates of local news and a few sips of coffee, I launched into my questions about the beginnings of the Sitka Center and the establishment of the Cascade Head Scenic Research Area. Frank, a natural raconteur, soon let loose a detailed story of those early efforts to conserve Cascade Head: an intriguing tale of diverse visions, entrepreneurial pitches, fights between neighbors, political ploys, and complicated financial mechanisms. I scribbled notes as fast as I could while my coffee got cold. "Man, it was really heady times," Frank said. "There was all this back-to-the-land stuff, people were starting little farms . . . well *you* know, you were there too!"

I smiled at Frank's use of the word "heady"—it's not used all that much these days. And I knew exactly what he meant. The dictionary says "heady" means "tending to intoxicate or make giddy or elated," and gives an example of use in the phrase "a heady wine." It wasn't only a cresting of environmental concern; it was also the endgame of the Vietnam War. We were all giddy and excited with the idea that we could make a difference, stop the war, change the world, save the Earth. That was the big social context of the hike on Cascade Head in 1973.

The events that led to me sitting in a cozy living room in Morley House on the Sitka Center campus chatting with Frank reach back to those years. When Senator Packwood climbed the salt-spray meadows of Cascade Head, I was in the final months of my senior year in college, graduating with a degree in human biology, a program that had been started by Paul Ehrlich, the ecologist and author of *The Population Bomb*. But my karma for being at Cascade Head reaches much farther back; a mycelium of memory binds me to this place.

I think it started at around five years old with my grandfather, Harry Sweeney, who took me to the tide pools of Haystack Rock at Cannon Beach when we visited him in Oregon, as we did most summers as I was growing up in landlocked New Mexico. Those tide pools gave a glimpse into another world. It seemed to me a huge, beautiful, nonhuman world of scuttling hermit crabs, snails, camouflaged fish, flower-like anemones, and purple, orange, and brown seastars. It was like a trip to another planet. I think that was the seed of my sense that the world is a wide and wondrous place, way beyond our grasp of it or influence on it, and that everything is connected to everything else. Sometime during my early high school years my relatives in Portland sent me an announcement about a summer marine science camp called Camp Arago, sponsored by the Oregon Museum of Science and Industry, in Charleston, on Coos Bay. I applied, was accepted, corresponded with the director, Bill Arbus, and asked him what I should do to get ready for camp. He suggested reading *Between Pacific Tides*, an ecologically oriented field guide to the Pacific intertidal zone written by Edward F. Ricketts. The librarian at the Los Alamos Library dutifully ordered it through interlibrary loan. I suspect she was perplexed about why I would want such an obscure title, although I don't remember her treating me like I was crazy. The book arrived from somewhere, and I read it from cover to cover. The poetic Latin scientific names of dozens of intertidal creatures I'd never seen embedded themselves in my brain: *Diaulula sandiegensis, Hermissenda crassicornis, Pisaster ochraceus, Pollicipes polymerus, Lottia digitalis, Anthopleura xanthogrammica, Henricia leviuscula, Strongylocentrotus purpuratus*. And I saw them all in the bright cold tide pools at Camp Arago the following summer. I was hooked.

The backstory Frank told me over coffee that morning about the "heady times" at Cascade Head made it clear that establishing the Cascade Head Scenic Research Area had been quite a local fight, sometimes pitting neighbor against neighbor: "There were some bad feelings then. Oh god, it wasn't pretty!" There were even death threats made against the anti-development group, Frank said.

He first sketched a brief history of the area, leading up to 1974. The headland meadows on the southern flank of Cascade Head had been acquired as a cattle ranch by James Savage in 1898, just a few years after the Dawes Act of 1887 opened up the lands in Tillamook County to non-Indian settlers. In 1912, Savage's daughter married Lloyd Gentry, and her father gave them the ranch as a wedding present. In 1965, Mike Lowell, a businessman with an MBA from Stanford, was trying to buy the 450-acre Cascade Head Ranch from the Gentrys. He wanted to develop an environmentally friendly "green" community modeled on Sea Ranch in California—but he needed money for the down payment and the road system. To protect the natural values of the area, and especially the headland with its scenic salt-spray meadows, he approached The Nature Conservancy (TNC) and offered them 270 acres for $50,000.

At first, TNC wasn't interested. So, Malcolm Montague, a Portland lawyer with a house on the estuary below the ranch, who, like Lowell, wanted to control and slow down the development of the area, came up with a plan. He approached Harold Hirsch, the wealthy owner of White Stag, a skiwear and outdoor clothing company based in Portland, with the problem. Hirsch, according to Frank, advised them to "plat it." So they had the whole area surveyed and produced a map showing the headland covered with lots, which Hirsch took to TNC. He wrote them a check for $5,000 and suggested they raise the other $45,000 to acquire the Cascade Head meadows as a nature preserve before they were covered in houses. That persuaded them to buy; TNC acquired the land in 1966 and established the Cascade Head Preserve. Lowell had his down payment and money for the road system at Cascade Head Ranch.

Meanwhile, some other local residents with land along the estuary below Cascade Head were pressing forward with development plans of their own. The Calkins family owned a boat works in a cove there and, starting in the 1920s, would rent little plots in the woods nearby to families from Salem and Portland who came in the summer to camp and enjoy the coast. When the city of Salem got rid of its streetcar system, Calkins bought the old cars and moved them to the area for people to rent as vacation cabins. There was a store and even a post office near the boat works, and the area seemed poised for a boom of tourism and development until the Depression dampened the process temporarily. But by the mid-1960s, big plans were afoot to develop a dense neighborhood of second homes on the Calkin's property along the estuary. Blueprints now displayed in the Cascade Head Ranch River House near the old boat works show hundreds of lots platted in a development called Three Rocks, Inc.

Malcolm Montague was not happy with what he saw coming to the place where he had spent so much time with his family, and which he loved so much. He wanted to stop Three Rocks, Inc., and used his personal connections with Harold Hirsch to approach Senator Packwood to start the political process that eventually led to the Cascade Head Scenic Research Area (CHSRA)—"chess-ra," as some people now call it, turning its abbreviation into an acronym.

"By then, there was quite a bit of fuel behind this thing," Frank told me, "but who was going to do it? They went to the Forest Service, and the Forest Service said 'Gee, well, we really don't know if we can . . . We're really busy.' So, Packwood threatened that maybe the National Park Service would do it—basically we were talking about putting together a park here—and right away the Forest Service jumped and said 'We'll do it!'"

Frank and Jane Boyden were among the conservationists conspiring with Montague, Lowell, Hirsch, and Packwood to protect the area from unchecked development. They'd arrived on the scene in the early 1970s to found an art-and-nature camp for kids at a site on Lowell's Cascade Head Ranch; in a few years the camp evolved into the Sitka Center for Art and Ecology.

"The CHSRA legislation sneaked through, and the locals went ballistic," said Stephen Dow Beckham, a prominent Oregon historian who at the time was on the Sitka Center board, when I interviewed him one morning at his home in Neskowin's South Beach neighborhood. It was "a year of high tensions and uncertainty about how the Forest Service would administer the area and the private lands within the designated boundaries." The Sitka Center hosted a series of meetings with local landowners and residents, and those gatherings "seemed to do a lot to defuse the fears, especially when property owners began to understand that the environment they valued would be preserved in the longer term," Beckham told me. "The gatherings had modest funding from the Oregon Committee for the Humanities. If memory serves me, the theme was 'Man and the Land.' Today that sounds sexist, but in the 1970s it was an effort to spark public discussion about environmental issues in Oregon. Cascade Head was a good case in point."

3 *The Midnight Forest and the Spruce Goose*

It would be desirable to set up a natural area in the northwest corner of the experimental forest as a "museum piece" of the virgin spruce and hemlock forest.

—Thornton Munger, letter to Fremont McComb, May 1940

Landscape ecology is a branch of ecology that seeks to understand ecological patterns and processes at relatively large landscape scales, and the history of the dynamic forces that created them. It has generally focused mainly on the natural forces that shape ecological landscapes. But dynamic political and economic forces have shaped the forests of Cascade Head just as surely as fire and wind have. Understanding how we have imposed our ideas, values, and worldviews on the landscape could perhaps be called "landscape ideology." At Cascade Head, we can look back in time and see how history and politics have shaped its forests.

One important example of human values imposed on this landscape is its inclusion, in 1907, in one of what came to be called the "midnight forests." To begin that story in a few fairly easy steps—with apologies if the steps in this storyline seem erratic at first—we should get to know a little bit about Charles Sprague Sargent, Harvard botanist, mentor and friend of John Muir, and leader of the National Forest Commission of 1896–1897. Sargent had established a journal called *Garden and Forest* in 1888, which called for "the reservation of all forested public lands from sale and settlement and their protection, along with that of the national parks, from thieves and vandals by the US military."

Although this may sound a bit harsh and radical to our ears, Sargent wasn't really going out on a limb. Two years earlier, in 1886, the US Army had been sent to Yellowstone National Park to protect it. Created in 1872, Yellowstone was the world's first national park. Established to preserve an ecosystem in its natural state, it was a landmark in landscape ideology. But within a decade after the creation of the park, it was clear that the Yellowstone ecosystem was in danger. Although its creation was at first urged by Montanans, the park was surrounded by a class of old frontiersmen, hunters, trappers, and miners who had no respect whatever for the rules and regulations established by the secretary of the interior in faraway Washington, DC. In 1883, Congress was becoming worried about the situation, and provided that "The Secretary of War, upon the request of the Secretary of the Interior, is hereby authorized and directed to make the necessary details of troops to prevent trespassers or intruders from entering the park for the purpose of destroying the game or objects of curiosity therein." So, in 1886, the management of Yellowstone National Park was turned over to the US Army. Soldiers chased poachers and fought forest fires, some of which were set by hunters to drive bison, elk, and other game out of the park so they could shoot them. The army was so effective in protecting Yellowstone that in 1891, when it became clear that Yosemite and two other parks in California were threatened by local development and mismanagement, troops were also sent there. By 1916, army "rangers" posted in our national parks had evolved into a new kind of paramilitary protective force, and the National Park Service was established.

A few years after Sargent's call to arms, the Forest Reserve Act of 1891 authorized the president to designate "forest reserves" on public lands in the West, to be managed by the Department of the Interior. President Harrison designated fewer than a dozen reserves by the time he left office in 1893. Sargent, along with other notable scientists, pressed the National Academy of Sciences to study the situation of western forests with an eye to designating more forest reserves and protecting them for the long-term public good. In 1896, Congress authorized the National Forest Commission, and President Grover Cleveland appointed Sargent as its head.

Sargent invited John Muir to join the commission as an adviser on their grand tour of western forests in the summer of 1896, although Muir was not an official member. Gifford Pinchot, a thirty-one-year-old French-trained American forester who would later become the first chief of the US Forest Service, was an official commissioner. Sargent and Pinchot didn't see eye to eye. Inspired by Muir's writing and his defense of Yosemite, Sargent advocated preserving the forests in a state of "wilderness," while Pinchot, with his European forestry training, advocated conserving forests for sustainable uses, including timber production. But Muir and Pinchot, who had met each other before, got along well. Although of quite different ages, they were the rebellious bad boys of the expedition, going off to sleep under the stars while the others huddled in their tents.

On July 26, 1896, Muir was interviewed in Portland by the *Morning Oregonian* during a stop on their tour of forests. What he said seems to call into question the common interpretation that Pinchot and Muir held opposing views about forest management. Muir said,

> Something must be done to preserve and perpetuate the forests, for the timber must ultimately be used. The forest must be able to yield a perennial supply of timber, without being destroyed or injuriously affecting the rainfall, thus securing all the benefits of a forest, and at the same time a good supply of timber. The establishment of national parks and [forest] reserves is only the beginning of the work necessary to secure these lands.

This sounds like a rather utilitarian view, not much different than what Pinchot might have said. In the end, according to Donald Worster, in his masterful biography of Muir, "the commission proposed a system that would both earn a profit and protect vital watersheds and timber supply, that would include Sargent's army guardians and Pinchot's civilian managers. Soldiers, the report suggested, should take charge of the public forests temporarily."

Acting on the report of the National Forest Commission, President Cleveland had declared thirteen new forest reserves, to much outcry from some western states. In 1901, Teddy Roosevelt became president after William McKinley was assassinated, and he won election to his own term in 1904. Under Roosevelt, the Transfer Act of 1905 moved the management of forest reserves from the Department of the Interior to the newly created US Forest Service in the Department of Agriculture. Roosevelt appointed Gifford Pinchot as its first chief forester. President Roosevelt continued to create new forest reserves and expand old ones; between 1904 and 1906, ten new reserves were established in Oregon alone.

In response, Oregon senator Charles W. Fulton introduced an amendment to the Agriculture Appropriations Act of 1907, which stated, "Hereafter no forest reserve shall be created, nor shall any addition be made to one heretofore created, within the limits of the States of Oregon, Washington, Idaho, Montana, Colorado or Wyoming." The amendment also changed the name of the forest reserves to "national forests," to indicate that they were to be used, not preserved. You can almost hear him muttering, "We can't pay our taxes with salal bushes!" and "We don't need Big Brother to advise and regulate."

The agriculture bill passed in February 1907, and Roosevelt knew he did not have the support to veto the broader act, which affected all of American agriculture. He had until noon on March 4, 1907, to sign it. So, he huddled with Gifford Pinchot, reviewed the notes from the National Forest Commission's reconnaissance, and Pinchot's staff got to work drawing up maps and proclamations for twenty-one new forest reserves and expansions of eleven others, with an area of more than sixteen million acres. When the paperwork was ready, Roosevelt signed it all by midnight on March 2, 1907, and the next day signed the Agriculture Appropriations Act with Senator Fulton's now impotent amendment—its only effect being to turn "forest reserves" into "national forests." The new and expanded national forests created by Roosevelt and Pinchot were dubbed the "midnight forests." According to Forest Service historian Harold Steen, "Roosevelt later gleefully recalled how opposing interests 'turned handsprings in their wrath' over the setting aside of these 'midnight reserves.'"

Five new national forests were proclaimed in Oregon: Blue Mountains, Coquille, Imnaha, Tillamook, and Umpqua National Forests. Approximately a year later, in July 1908, the Siuslaw National Forest was created from parts of the Tillamook and Umpqua National Forests.

So, Cascade Head, in the Siuslaw, lies within a "midnight forest," the outcome of a struggle between a national and a more local vision of who should manage our forests and how they should be used. In resisting the tendency for local overexploitation and imposing a wider view of the common good, Pinchot and Roosevelt set the stage, so to speak, for the drama that established the Cascade Head Scenic Research Area almost seven decades later, a drama with a parallel tension between local and national values.

~

Thornton T. Munger (1883–1975) arrived in Oregon in 1908, one year after the creation of the midnight forests, having completed a master of forestry degree at the Yale Forest School earlier that year. He was one year ahead of Aldo Leopold at the school, and they must have known each other. Munger was the first director of the US Forest Service's Pacific Northwest Forest Experiment Station, from 1924 to 1938. When he took that job, the forest industry was racing ahead of forest science, cutting "a wide swath through what it considered a 'one-time crop' without regard for the future," according to the *Oregon Encyclopedia*.

"We have no time now for research for research's sake," Munger said, and he focused the experiment station's research toward "the important timber belts, where the extensive lumbering operations lie and where there are great areas whose future may be either devastation or reforestation. The selection of projects will depend on their economic importance." In a speech to the Pacific Logging Congress in Portland in 1924, Munger said, "It may be presumptuous to say that man can improve on Nature, but he certainly can if he goes about it." He also stated that "under proper management man can produce more wood per acre than Nature has in the wild stands."

What was Munger thinking? Was he only a hopeful, naïve forester of his day, lacking in ecological humility, who had some notion that humans could improve on natural processes? The explanation is probably a bit different. I think it probably has to do with Munger's understanding of the ecological dynamics of Pacific Northwest forests, but with the value system of the Forest Service of his day superimposed on that understanding.

Aldo Leopold, Munger's fellow student at the Yale Forest School, was a master at reading the ecological history of a landscape. Leopold developed that skill during his first assignment with the US Forest Service in Arizona and honed it at his property on the Wisconsin River in the last decades of his life, while writing *A Sand County Almanac.* Munger probably had a similar eye for reading ecological landscapes. He knew that parts of Cascade Head had been burned in about 1845 in a widespread historic fire called the Nestucca Fire, which had reset the ecological clock there and restarted a process that ecologists call "succession." He wrote an article about this, "Out of the Ashes of Nestucca," published in 1944 in *American Forests.* He was aware that the young forest of Sitka spruce, western hemlock, and Douglas-fir that had sprung up naturally after the fire was in its exuberant youth, growing as fast as forests can grow. As Munger saw it, the challenge for forest research was to "discover the ecologically and economically viable silvicultural systems that would convert the stagnant or slow-growing, old-growth forests east and west of the Cascade Range into faster-growing and more-productive second-growth forests." Perhaps clear-cutting forests could do what fire did, turning slow-growing old forests into fast-growing young ones?

Munger became the director of the Northwest Forest Experiment Station only a half-dozen years after the end of World War I, a war in which military aircraft debuted. Bi-winged and tri-winged planes made of wood and fabric fought in daring "dogfights" and introduced a new technology and element to war. Spruce wood, it turned out, had properties of flexibility, tensile strength, and especially "torsional strength" (resistance to bending) that put it at the top of the list for many structural components of planes—wing beams, struts, longerons, ribs, and landing gear—that would later be replaced by steel. So, it shouldn't come as a surprise that in the 1930s, with World War I in the rearview mirror and ominous developments in Germany hinting at conflict to come, the US Forest Service would be asked how a lot of spruce wood could be produced in case the country should need it. A Forest Service report titled "Aircraft Woods: Their Properties, Selection, and Characteristics" stated that "Sitka spruce, because of its large size, availability, and proportion of clear stock in suitable sizes . . . is the chief source of supply." There was no better place within the National Forest System to study how to grow Sitka spruce big and fast than at Cascade Head. In that context, the Cascade Head Experimental Forest was established in 1934—under the supervision of Thornton Munger.

The Spruce Goose is the nickname for a giant plane, the Hughes H-4 Hercules, that was designed to transport troops and materiel across the Atlantic in World War II to avoid the risk from German submarines. Because of wartime restrictions on metals, and weight considerations, it was built almost entirely of wood. Although birch was the major component of the plywood sheathing of the plane, its structural members were mostly spruce—probably Sitka spruce—and hence its nickname. With eight engines, it had the largest wingspan of any aircraft that had ever flown, and it was built as a flying boat, able to land and take off from water. Only one prototype was produced, and it was not completed in time to be used in the war. Perhaps by some cosmic coincidence, the original Spruce Goose has finally come to roost at the Evergreen Aviation and Space Museum near McMinnville, Oregon, not far from where it hatched in a midnight forest, where it is now a main attraction.

~

The historic headquarters of the Cascade Head Experimental Forest near Otis, Oregon, is a small campus of two 1930s buildings surrounded by a big lawn just off Old Scenic Highway 101, the main coastal road until the modern, straightened version was completed in 1961. One building was originally the director's house, with its bright white clapboard siding, forest green shutters, brown shingle roof, and red brick chimney, now occupied by a caretaker who watches over the place. The other, in matching architectural style, was the former office and living quarters for employees on temporary assignments, with an attached garage and shop. Ray Touchstone, a retired local schoolteacher who is now the caretaker, invited me in on a rainy afternoon to show me the original wooden filing cabinets, still sitting in a front room of the old house. The files, musty and organized just as they were in that old-fashioned era, went back to the mid-1930s. I poked through the ones that seemed relevant to my investigation of the history of the place. In a section labeled "Research" I found some letters from 1940 and early 1941 that opened a window onto the creation of the Neskowin Crest Research Natural Area within the Cascade Head Experimental Forest.

The first letter was on letterhead of the director of the US Forest Service Pacific Northwest Forest and Range Experiment Station, dated May 11, 1940, to Fremont McComb, a forester at the experimental forest:

> Dear McComb:
> Last year when Leo and I were at Neskowin Crest we decided that it would be desirable to set up a natural area in the northwest corner of the experimental forest as a "museum piece" of the virgin spruce and hemlock forest. Though some of this area is pretty close to the ocean, some is in protected coves and is quite typical of the spruce-hemlock coastal type. . . . I should like to have you spend some time in the area concerned checking the 1934 type map, correcting it if necessary, and adding detail where needed. Before deciding upon the most desirable boundaries for this natural area, we would like to know quite accurately the extent and location of the following types; old-growth spruce, old-growth hemlock, second-growth spruce, second-growth hemlock, and of course of such other distinctive types as grass land, brush land, etc. It may be desirable to add still another type where the virgin timber is scattered and there is an intimate mixture of old-growth and second-growth trees.

The letter was from Thornton Munger. According to Dr. Jerry Franklin, the eminent forest ecologist who was a later research director at Cascade Head, Munger was very active in establishing research natural areas and was friend and mentor to Leo Isaac—the Leo mentioned in the letter.

Mr. McComb's response to the request was a report, a bit farther back in the file, dated January 20, 1941, and titled "The Proposed Neskowin Crest Natural Area." To me it was telling and touching in its own technical but personal way. I imagine the author and his readers feeling an ominous shadow looming over the landscape; unknown to them at the time, Pearl Harbor was less than a year away. McComb wrote,

> Sitka spruce and western hemlock comprise nearly all the forest cover. There are roughly two age classes—old growth, where the trees are probably 250 years old and uneven aged and second growth where the trees are uniformly about 90 years old. In places the old growth prevails over a considerable area, especially in the western part of Section 2. There are splendid specimens of both

> spruce and hemlock, ranging up to over five feet in diameter for the former and four feet for the latter. . . . Some of the hemlock is very fine with tall clean cylindrical boles, but much is decrepit as usual in old uneven-age stands.

Here we have mostly a technical description, with a few hints where personal and professional values of the time muddle the wording; we see "splendid specimens," but "decrepit . . . old uneven-age stands" describing the same forest. And then comes an interesting paragraph in the report:

> There are patches within the area which are fine examples of a type which is fast disappearing in Oregon and Washington, and is rare within national forest boundaries, virgin old growth spruce-hemlock, in which hemlock is gaining the ascendancy by the gradual dropping out of the spruce and the preponderance of the hemlock in the reproduction.

McComb was reading and projecting forest succession here, and arguing in a quiet, understated way that it was important to protect areas like this because they could help forest ecologists understand ecological succession.

A paragraph headed Plan of Management in McComb's report said, "This area is set aside for the purpose of maintaining a tract of Sitka spruce and western hemlock type in its natural condition." This was to be the "museum piece" of virgin spruce and hemlock forest that Thornton Munger had wanted to establish.

~

Another paragraph in McComb's report, under the heading "Transportation," piqued my curiosity. It reported that "a passable trail, formerly a cart trail, passes through the western end of this area going from Neskowin South Beach to Neskowin Crest, and on to the terminus of the Ridge Road. It is likely to be used considerably by tourists, but this will have no detrimental effect on the natural area."

I had heard about this old trail because I'd been attending meetings convened by the Forest Service's Hebo Ranger District to gather local input about impending management decisions from a group of people representing various organizations and landowners with an interest in the Cascade Head Scenic Research Area. The group was wrestling with issues of access to the area, and trails, parking at trailheads, overuse of trails, and private landowners' concerns about those things. The trailhead for the old trail to Hart's Cove was now on private property, closed to public access, within the residential development at Neskowin's South Beach. The Forest Service considers the trail officially "closed," but it is known and used by residents of South Beach and their friends. This had come up as one of many controversies that needed to be sorted out by the group. When I saw McComb's report, I knew that was the trail he was talking about, and I wanted to hike it and see the "museum piece" forest of the Neskowin Crest Research Natural Area through which it led.

I had become acquainted with a longtime property owner at South Beach, and he was happy to accommodate—even encourage—my interest in the South-Beach-to-Hart's-Cove trail. A few days after Christmas, I punched the code he'd given me into the keypad at the gate on South Beach Road; it slid open, and I drove across the bridge over Neskowin Creek to his house. He guided me to the trailhead, where I laced up my boots and started on my way.

The trail climbed up above the twin water tanks of the South Beach water utility, and then stayed mostly on contour as it arced along the valley of a small unnamed stream, crossing it and turning back coastward in a "V" along a gradual climb to the next ridge south. Ah! Here were the "decrepit . . . old uneven-age stands" McComb had written about in his 1941 report to Thornton Munger. Giant spruces with wind-broken tops had here and there grown new vertical trunks from branches below the break. The horizontal branches had grown thick to support the weight of the new sky-striving vertical trunks, and the wide moss mats that cover them are the preferred nesting sites of marbled murrelets.

The trail was clearly used, but not heavily. Delighted by this chance to walk in true old forest, I didn't mind that it wasn't perfectly maintained. I had to climb over, under, or around several fallen trees, but it appeared that other treefalls had been chainsawed open within the last decade. The top of every log was a nursery of tiny trees, a forest in miniature. The forest floor was a thick green moss mat.

In about an hour, the faint trail emerged from the trees into a sloping meadow with a grand, nearly 180-degree view of the ocean. The meadow is being pinched from the edges by a thicket of salmonberry—part of the successional process here—whose closely spaced and spiny stems create a barrier to passage that only the elk dare to tackle. I'll take an assault through salal to one through salmonberry any day.

This meadow figures in the history of the area detailed in a 1975 report titled *Cascade Head and the Salmon River Estuary: A History of Indian and White Settlement,* which was prepared for the Forest Service by Stephen Dow Beckham, the historian. That study was required by the Antiquities Act and the National Historic Preservation Act when Cascade Head was designated a Scenic Research Area. According to Beckham, the meadow and surrounding forest belonged to James Taggart and his wife Jessie, who acquired it in 1896 through the Homestead Act. They sold it in 1916 to Charles Hart for $2,500, and he built a log cabin on the eastern edge of the meadow, where he grew potatoes that he hauled over a cart track to Neskowin, a couple of miles north. According to Beckham, who interviewed Hart, then ninety-five years old, in 1975,

> He sold his potatoes, usually for $1.00 per sack, to tourists camping at Neskowin. Hart also taught several sessions of school at Otis and walked each day from the cape over Cascade Head to the Salmon River to teach and returned the same night to his cabin. Hart recalled that as a young man his gait was 5½ miles per hour. "Now," he said, "I only walk five miles an hour!"

If Hart's potato patch was written in the grassland and salmonberry here, I couldn't read its exact location. While I ate lunch, I scanned for spouting whales through binoculars before packing up and plunging into the forest again, back north to Neskowin.

4 *Old Orchard and the Biosphere*

> *Nature is not a place to visit, it is* home*—and within that home territory there are more familiar and less familiar places.*
>
> —Gary Snyder, *The Practice of the Wild*, 1990

As I walked westward toward Cascade Head across what my Forest Service contacts called the Reference Marsh, following elk trails through the tawny, knee-high grasses and rushes, I heard voices from the woods to the south. Through a break in the alders fringing the marsh I could see with my binoculars an older couple, picking apples from an unkempt tree tucked into a clearing. The man had a long pole to knock the fruit down from the high branches, the woman a big sack to put the apples in as she stooped stiffly to pick them up.

On my return from the reconnaissance of the marsh, I followed up the west bank of Rowdy Creek, past where the Tamara Quays Trailer Park had been removed in 2009 and the marsh restored by removing dikes. The elk had been here too, along the muddy track. The marsh was healing. I clambered through the blackberries, an almost impenetrable hedge along the marsh edge, like the coils of razor wire along a World War I trench keeping the enemy away. But there were a few trails hacked through, by fishermen I assumed, and I angled left on an old road, looking for the old orchard from the landward side. As I rounded the alders into the clearing, a coyote loped off. Apple harvesting too, I supposed. There were at least a dozen trees, each a different variety it seemed, probably some old-time heirlooms. Most of the fruit was blemished in some way, though firm and ripe. Some were tiny—the trees hadn't been pruned or tended in a long time.

People used to live here, obviously: the Frasers, who grazed their cows on the Reference Marsh and cut hay from the high places along the river, where it had deposited sediment and built up a natural levee that stayed above the tide longer than most of the rest of the marsh. But the place is rewilding now, slowly, with elk and coyotes coming back, and clearings closing to alders and shore pine, blackberries and spruce. In a half century the apples will all be gone.

I cut one of the bigger apples in half, partly to make sure there weren't worms in the core, and was struck by the contrast between the white flesh and the delicate red skin. My mind jumped to an analogy I used a lot when I taught ecology: the "biosphere," the sphere where life exists, is about as thick relative to the whole Earth as the skin of an apple is to the apple. In other words, thin.

The Earth's biosphere is where the nonliving parts of our round planet, the lithosphere, hydrosphere, and atmosphere, meet and greet and create the mix-up in which life evolved and exists. Each of these "spheres" is a dynamic system. The lithosphere vomits volcanoes, deposits sediments, faults, quakes, and pushes up mountains. The hydrosphere dumps downpours, floods valleys, carves canyons, and attacks coasts. The atmosphere flattens forests with windstorms, hammers coastlines with hurricanes, girds the latitudes with its wind belts, and traps solar heat under its blanket of carbon dioxide. But the biosphere is more dynamic still, with its food webs, nutrient cycles, symbioses, ecosystems, and evolution. For us, the biosphere is our one and only home here on this far-flung lonely island Earth circling an undistinguished star in an outer band of the Milky Way galaxy. And it's thin, very thin.

The thickness of the biosphere? Well, likely from the depths of the deepest oceans, maybe six miles deep, to at or a little above the tops of the tallest mountains, maybe six miles high. Twelve miles thick, compared with the diameter of the Earth, maybe eight thousand miles—the living skin of the Earth apple is, then, not much more than one-thousandth as thick as the planet is wide. If you look at the famous Apollo 17 photo of Earth, sometimes called "the blue marble" photo, you can barely see the thin skin of atmosphere around the edges. There is just something about that thought that gives almost anyone pause. Thin seems to imply fragile, or at least raise a question about fragility.

Our sister planet, Mars, another rocky planet, has a lithosphere of course. It may once also have had a hydrosphere and a thicker atmosphere, but if so, those are mostly gone now. We don't yet know why. And, if it had those three creative nonliving components, some think it may also have had its own biosphere, although convincing evidence awaits future exploration. But at the very least Mars raises a question about the fragility of the systems of the "spheres."

The thinness of our biosphere is one reason that one of its species, having developed a technological culture, could significantly alter it. Our discovery of how to burn fossil fuels less than two centuries ago has caused changes in both the atmosphere and the hydrosphere that are irreversible on time scales ten or a hundred times as long. Or maybe never, if we have triggered some tipping point, like whatever happened on Mars. We have now begun to acidify the oceans, and probably to change their patterns of circulating currents that redistribute the heat and cold of the planet. That effect, caused by increasing concentrations of carbon dioxide in the atmosphere, is also changing the average temperatures of the land and oceans, conditions on which our current societies and economies depend, with wholly uncertain and unpredictable consequences.

Our biosphere may be the only place in the universe where the conditions were just right and life developed. We don't know for sure—although many biologists think life may not be that unusual in such a vast old universe. But, since we don't know for sure, we might as well assume that our biosphere is *the* biosphere, the one and only—Spaceship Earth, as it was often referred to during those heady days of the rise in environmental consciousness that began in earnest in the 1960s.

~

The term "biosphere" was first used in something like its modern sense by the Austrian geologist Eduard Suess, in his book *Das Antlitz der Erde*, or *The Face of the Earth*, published in 1885. The term and concept were promoted by Ukrainian biogeochemist Vladimir Vernadsky, in a 1926 book, *The Bio-*

sphere, which was translated from Russian to French in 1929, and soon after to English. Ecologist Frank Golley describes Vernadsky's book as "a scientific expression of a global system of man and nature, which was an antidote to the virulent nationalism that was being expressed at the time, especially in Europe." As such, the biosphere concept can be seen as resistance to the nationalism and fascism that were beginning to cast their dark shadow over Europe in the 1920s.

In 1968, with the Cold War continuing, the Vietnam War under way, and concern about environmental threats exploding, the United Nations Educational, Scientific and Cultural Organization (UNESCO) organized the Biosphere Conference in Paris, using the word "biosphere" for the first time in international deliberations. In a UNESCO retrospective on the legacy of the conference published in 1993, the authors wrote that "the single most original feature of the Biosphere Conference however was to have firmly declared that the utilization and the conservation of our land and water resources should go hand in hand rather than in opposition, and that interdisciplinary approaches should be promoted to achieve this aim." The biosphere concept itself was an act of resistance against the idea that biodiversity conservation and human development are incompatible or contradictory.

A few years before the Biosphere Conference, in 1964, the International Council of Scientific Unions had launched a ten-year program of international cooperation called the International Biological Program, or IBP, modeled on the success of the International Geophysical Year of 1957–1958, to better understand the functioning of ecosystems at large scales. Science seemed to be a tool for chipping away at geopolitical and ideological walls. Following the Biosphere Conference, UNESCO established the Man and the Biosphere Programme in 1971. It combined the environment-and-development perspective of the conference and the large-scale, long-term, ecosystem ecology research of the IBP, and sought to establish a network of places, distributed around the diverse ecosystems of the biosphere, where we can monitor, study, assess, and respond to the changes that humans are causing.

Two big events in this story occurred in 1972. One, the UN Conference on the Human Environment, held in Stockholm, Sweden, continued deliberations about how to save the biosphere. The second, the Moscow Summit between President Richard Nixon and Soviet general secretary Leonid Brezhnev, was a major step toward Cold War détente. Following the summit, US and Soviet scientists were tasked with finding ways to work together on issues of mutual interest. The ecosystem research already mounted under the IBP and the proposal for an international network of biosphere reserves seemed to be a place to start. It just so happened that, in 1973, forest ecologist Jerry Franklin, who had spent the summer of 1958 at the Cascade Head Experimental Forest as a forestry student trainee from Oregon State University and then had risen through the ranks of the Forest Service, went to Washington, DC, to serve at the National Science Foundation. At NSF, Dr. Franklin was chosen to lead a US delegation charged with working with the Russians to establish biosphere reserves in the two countries. I wanted to hear the whole story from him.

~

I was on my way to Vancouver Island, British Columbia, to visit Mount Arrowsmith and Clayoquot Sound, two Canadian biosphere reserves, and since I was passing through the Seattle area, I had made an appointment to meet Jerry Franklin at 9:00 a.m. at the Volvo dealership in Bellevue, where his car was in for a checkup. He figured he would have a couple of hours, and we could have a quiet talk over coffee. I made it just in time, and Jerry was already there, instantly recognizable in his signature western-style broad-brimmed hat with a beaded hatband. The Volvo office was very quiet on this Monday morning and had free coffee. We had a wonderful, far-ranging conversation.

Franklin was at the National Science Foundation from 1973 to 1975, and went to Paris for the second meeting on biosphere reserves, where he and the delegates from the Soviet Union hit it off well, he said. Neither the United States nor the Soviet Union saw the benefit of just another program focused on protected areas. What they wanted was a concept that integrated science, management, and society in a broad vision. The process of selecting the first US biosphere reserves, which he led, was based on this concept.

Jerry saw Cascade Head as a natural place to establish a biosphere reserve. It built on a foundation of forest research reaching back to 1934, when it was made an experimental forest, and 1941, when the Neskowin Crest Research Natural Area was established. When Cascade Head became a Scenic Research Area in 1974, the management objective stated in the act of Congress that established it was, "To provide present and future generations with the use and enjoyment of certain ocean headlands, rivers, streams, estuaries, and forested areas, to insure the protection and encourage the study of significant areas for research and scientific purposes, and to promote a more sensitive relationship between man and his adjacent environment." As such, the goals for CHSRA meshed well with the objectives of the UNESCO Man and the Biosphere Programme, just at the time the first US biosphere reserves were being selected.

Several of the first crop of twenty-nine US biosphere reserves were centered on Forest Service experimental forests—the fingerprints of Jerry Franklin's influence on their selection. But Jerry complained to me that the Forest Service never really "adopted" biosphere reserves, whereas the National Park Service, whose national parks were the loci of many other early US biosphere reserves, took to them much more readily. On the other hand, Jerry said, no one in the Forest Service told them they couldn't have biosphere reserves; "It was below everybody's radar."

Although each biosphere reserve in the Man and the Biosphere (MAB) Programme was supposed to conduct a periodic review and update its plans every ten years, almost none of the US biosphere reserves did so, until UNESCO began to increase the pressure, beginning around 2014. At Cascade Head, a team from the Forest Service and the Salmon Drift Creek Watershed Council, with the help of other key partners, did conduct a periodic review—titled, prosaically, "Periodic Review for Biosphere Reserve"—which was approved by UNESCO in 2016.

When the Cascade Head Biosphere Reserve was established in 1976, it consisted only of the Cascade Head Experimental Forest and Cascade Head Scenic Research Area, with a total area of about 21,500 acres. By the time of the 2016 periodic review, UNESCO was promoting a standard model of zonation within biosphere reserves that included "core" protected zones, "buffer" zones with minimal human disturbance surrounding those core areas, and "transition" zones of cities, farms, and other human activities around the buffer zones. The idea was to protect examples of undisturbed ecosystems in the midst of a human-modified, and often human-dominated, landscape—a worthy idea, but hard to implement in a simple way almost anywhere in the world. At Cascade Head, the periodic review team updated and analyzed the land use and land management situation and redefined the zones of the biosphere reserve. The entire seventy-five-square-mile watershed of the Salmon River was included in the overall boundaries, as were the new Cascade Head Marine Reserve and adjacent Marine Protected Areas. The "Periodic Review" report explained that the evolution of watershed-scale conservation efforts and a recognition of the important linkages between ocean and land argued for "a more integrated reserve area that includes a broader array of ecological and economic interests." In updating the zonation of the Cascade Head Biosphere Reserve, the Neskowin Crest Research Natural Area, the Reference Marsh and other restored salt marshes of the Salmon River estuary within the CHSRA, and the Cascade Head Marine Reserve became the core protected areas. Rather than using UNESCO's term "buffer zone," the "Periodic Review" adopted the term Zone of Managed Use, and included CHSRA and the Experimental Forest, TNC's Cascade Head Preserve, Westwind Stewardship Group land that is under a conservation easement, and the Cascade Head Marine Protected Areas, where fishing and other activities are less strictly regulated than in the Marine Reserve itself. The Salmon River watershed, and parts of Lincoln City to the south and Neskowin to the north of Cascade Head, were designated a Zone of Cooperation and Partnership. In all, the Cascade Head Biosphere Reserve now encompasses about 102,000 acres, or 160 square miles.

Although the US Forest Service is the official administrative point of contact with UNESCO for the Cascade Head Biosphere Reserve, what might be called the "stakeholder landscape" is very complicated. One category of stakeholders includes agencies with administrative and legal responsibilities, such as the US Forest Service, the Oregon Department of Fish and Wildlife, Salmon Drift Creek Watershed Council, the City of Lincoln City, Lincoln and Tillamook Counties, Oregon State Parks, the US Fish and Wildlife Service, the National Oceanic and Atmospheric Administration (NOAA), US Environmental Protection Agency, and the Confederated Tribes of Siletz Indians. Other landowners and land managers are also important stakeholders: The Nature Conservancy, Cascade Head Ranch, the Westwind Stewardship Group, the Sitka Center for Art and Ecology, and commercial timber companies such as Miami Corporation and Hancock Timber Resource Group. And then there are the academic and research institutions with important roles and interests, including Oregon State University and its Hatfield Marine Science Center. This complexity isn't unusual; every biosphere reserve has a similarly complex ownership and management context.

The idea behind the development of the UNESCO Man and the Biosphere Programme was that we need a network of places dedicated to monitoring and understanding the diverse ecosystems of the biosphere and developing models and strategies for maintaining or restoring their resilience while still meeting human social, cultural, and economic needs. In my work as an ecological consultant I've had the good fortune to spend time in thirty-four biosphere reserves in seventeen countries. Although each is unique, they all face similar challenges and provide lessons for all the others. Resistance, research, restoration, reconciliation, and resilience are common themes in the history and evolution of every one. I could see many of these themes on a memorable trip, in April 2017, to the Askania-Nova Biosphere Reserve in the steppe region of southeastern Ukraine, the home country of the founder of the biosphere concept, Vladimir Vernadsky.

That morning at seven o'clock, I'd met Alena Tarasova and Olga Denyshchyk, my Ukrainian team members, at the cavernous Soviet-era train station in Kyiv, Ukraine's capital city. We were conducting an analysis of biodiversity conservation in Ukraine for the US Agency for International Development. We boarded a train at seven-thirty, heading south down the Dnieper River—Ukraine's Mississippi, more or less. Our destination was Kherson, a city at the top of the Dnieper Delta—Ukraine's New Orleans, more or less. We arrived in the late afternoon and met with the head of the regional office of the Ministry of Ecology and Natural Resources, the agency in charge of all of the protected areas in the Kherson region. Leaving Kherson at dusk after the meeting, we drove across a big bridge spanning the Dnieper, and eastward into the dark in a Toyota RAV4 rented from Avis Ukraine. Our destination was the Askania-Nova Biosphere Reserve.

Our driver, Sergiy, sped along the deserted highway; I was too tired to do more than hope he missed the potholes. Finally, by nine o'clock, we were in Askania-Nova, checking into our rooms in the Hotel Kanna, where it seemed we were the only guests.

I left my bags in a huge, cold room and joined Alena and Olga in the cold restaurant. The proprietress was eager to get us fed and out so she could go home. After a typical Ukrainian dinner of soup, bread, salad, and meat, washed down with cold local beer, I was back in my room again. The electric heat had kicked on and warmed it up. I was thankful for that hydroelectric heat from the big dams on the Dnieper—or maybe from nuclear plants, of which Ukraine still has many, their inheritance from the Soviet era and its Chernobyl meltdown. The room had a 1950s "retro" feel that reminded me vaguely of the styles I grew up with. I especially liked the green carpet, and styling of mirrors, furniture, and lamps.

As I drifted off to sleep in the now-warm room, the carpet pattern became an image of a mosaic ecological landscape. I dreamed of a green harmony of multiple uses of the land, from farms to native grasslands grazed by . . . mammoths? saiga? sheep? If any of those beasts grazed through my dreams, I didn't remember them when my alarm jolted me awake at six-forty-five and sent me down to breakfast in the unheated restaurant. Breakfast was as prosaic as dinner had been—but, bless them, they had an espresso machine that made decent café americano!

After breakfast we drove to the campus-like headquarters of the Askania-Nova Biosphere Reserve, one of eight UNESCO biosphere reserves in Ukraine. Its history is deep and rich, as we learned from the director, Viktor Gavrilenko. He arrived in a tiny Soviet-era mini-jeep, appropriately olive drab in color, and whisked us off on a whirlwind three-hour tour of his empire. He has been director here for twenty-seven years.

From the headquarters buildings, Viktor drove us around the reserve while he related its history. From the mysterious 2,500-year-old Scythian stone statues that marked the salt-trading route from the Azov Sea across the waterless, trackless steppe to the German engineering that brought water up from an underground aquifer for a botanical garden and a giant sheep-ranching operation to a pioneering experiment in conserving vanishing steppe animals, this is a fascinating place.

The history of steppe conservation here reaches back more than a century, to Friedrich Falz-Fein (1863–1920), the eldest son of German settlers, who took over the family property near the German colony of Askania Nova after finishing his university studies in natural science at the University of Dorpat, in what is now Estonia. German immigration had been welcomed in southern Ukraine since the time of Catherine the Great, who ruled Russia from 1762 to 1796 and incorporated what is now Ukraine into her empire. Falz-Fein developed Askania Nova as a sheep ranch, said to have had nearly a million sheep and one hundred thousand shepherd dogs. The ranch also supplied horses to the Russian army. In 1883, Falz-Fein created an enclosure on native steppe for keeping wild steppe animals such as saiga antelope and, in 1899, brought a small herd of Przewalski's horses, which were even then nearly extinct in the wild, from the Mongolian steppe to Askania Nova. He knew that the steppe ecosystem, and especially its fauna, were threatened and rapidly disappearing.

In 1918, after the Russian Revolution, the Bolsheviks confiscated the property, and the Falz-Fein family fled to Germany—except for Friedrich's mother, who refused to leave and was executed by the Red Army. Under Stalin, Askania Nova was divided into agricultural cooperatives, but a core nature preserve was maintained as a State Steppe Reserve of the Ukrainian Soviet Socialist Republic. The main purpose of the reserve was to preserve and study an example of undisturbed steppe. In 1984, Askania-Nova was designated a biosphere reserve under the UNESCO Man and the Biosphere Programme, and Ukraine retained that status after the fall of the Soviet Union and its independence in 1991.

Victor drove to a gate that opened into a huge fenced area of native grassland, the enclosed park that Falz-Fein had established for endangered steppe animals. Although the habitat was natural, these fenced pastures had the feeling of a drive-in zoo and felt a little surreal. The grassland here was green with a flush of new spring grass. The pasture held a swirling herd of a few hundred saiga (*Saiga tatarica*), a steppe antelope with a pale coat that runs with a gait more like a dog than that of the bounding and springing African antelopes, their closest relatives. Saiga have a unique face, with elongated, drooping nostrils that are thought to warm frigid air in winter, cool hot air in summer, and filter dust on long migrations across the steppe. The species is considered critically endangered by the International Union for Conservation of Nature.

In another area, we watched a large herd of red deer (*Cervus elaphus*), the Eurasian ecological equivalent and sister species of the North American elk (*Cervus canadensis*). Viktor said that red deer would have been found here on the steppe before their local extinction. In North America, elk also ranged widely across the plains grasslands, before their local extinction there.

Viktor stopped suddenly at the edge of a pasture where bright yellow wild tulips, *Tulipa scythica*, whose scientific name honors the Scythians, scattered across the green grassland. This species is found only on the steppe of southeastern Ukraine and is listed in the Red Data Book of Ukraine as "disappearing," he said. Askania-Nova is one of its last holdouts.

Through another gate we entered another enclosure to meet the director's pride and joy, Askania-Nova's herd of Przewalski's horses. The taxonomic status of Przewalski's horse is equivocal. Wikipedia, citing authoritative scientific sources, says that "the taxonomic position of Przewalski's horse remains controversial and no consensus exists whether it is a full species (*Equus przewalskii*), a subspecies of the wild horse (*Equus ferus przewalskii*), or even a sub-population of the domestic horse (*Equus ferus caballus*)." The International Union for Conservation of Nature, which maintains the global Red List of threatened and endangered species, says that "current scientific review of the taxonomy of wild equids places Przewalski's Horse as a subspecies of the extinct *Equus ferus*" and notes that both genetic and morphological differences suggest significant evolutionary differences between Przewalski's horse and domestic horses. No matter what their evolutionary status, Viktor's familiar shouts and exhortations eventually persuaded the semi-tame herd to come over and greet us, and we could touch their wide faces and feel their un-horse-like, erect manes—more like those of zebras than of domestic horses.

Askania-Nova Biosphere Reserve is an enduring legacy of the conservation philosophy of Vasily V. Dokuchaev (1846–1903), an older contemporary of Falz-Fein. Dokuchaev was a pioneering Russian geologist and geographer who laid the foundations of soil science. Because of him, Russian names for soil types are common, such as "chernozem" for the black soils like those found on the Ukrainian steppe. Dokuchaev was instrumental in creating a unique Russian model of protected areas, called *zapovedniks*, best translated as "nature preserves" or "nature sanctuaries." Through the 1890s, Dokuchaev argued that setting aside areas of pristine natural ecosystems that can be compared with managed ecosystems, such as agricultural lands or managed forests, was ultimately important for economic development because they act as scientific controls to study how human actions affect ecological processes. *Zapovedniks* should be closed to all economic activities, he argued, and scientists should study their natural functioning.

In the United States, the *zapovednik*-like model of "nature preserves" exists to a certain extent in Forest Service Research Natural Areas, in our federal National Wildlife Refuge System, and in some preserves like those of The Nature Conservancy. But the philosophical foundations of nature conservation in the United States are in general based more on scenic, spiritual, and recreational values, growing out of the writings of Henry David Thoreau, John Muir, John Burroughs, and Teddy Roosevelt, in contrast to the utilitarian, scientific foundation of *zapovedniks*.

However, in the late 1930s, with the Dust Bowl disaster continuing, Aldo Leopold—a founding father of US conservation philosophy—understood the value of *zapovednik*-type nature reserves. In a 1938 essay titled "Engineering and Conservation," Leopold cited the research of John E. Weaver, a botanist, prairie ecologist, and professor at the University of Nebraska, and wrote, "While even the largest wilderness areas become partially deranged, it required only a few wild acres for J. E. Weaver to discover why the prairie flora is more drought-resistant than the agronomic flora which has supplanted it." The answer was that wild prairie plants had more complex, and more efficient, root systems, as Weaver discovered by studying the ecological processes in a small patch of undisturbed native prairie. Leopold expanded his vision of the value of preserving, studying, and learning from wild ecosystems in his 1939 essay "A Biotic View of the Land." He again cites Weaver, saying, "Professor Weaver proposes that we use prairie flowers to reflocculate the wasting soils of the dust bowl; who knows for what purpose cranes and condors, otters and grizzlies may some day be used."

Aldo Leopold would have enjoyed a conversation with Vasily Dokuchaev, Friedrich Falz-Fein, and Vladimir Vernadsky, and they would have loved to compare notes with him, John Weaver, and Jerry Franklin too, of course.

As I passed through the pastures of Askania-Nova, I could imagine us all crammed into Viktor Gavrilenko's little jeep, transported together in a time warp: listening to the calling skeins of cranes, watching the saiga swirl across the spring steppe abloom with yellow wild tulips, silently agreeing that this was another place where wild nature may be able to teach us—if we only listen—what we need to know.

Cascade Head and Askania-Nova are "thick places," where the multidimensional fabric of the biosphere doesn't already have a hole in it, or at least where it is not so worn and thin it is hard to know where to begin.

5 *Reference Marsh*

> *Salt marshes, scientists tell us, are microcosms of the world, held together by invisible strings of inscrutable complexity. By chemistries, geomorphologies, and biocontingencies that make the world one.*
>
> —Tim Traver, *Sippewissett*, 2006

The wind was still up this morning, as it had been for two days, but the sky was clear and the sun glorious. Thinking of the last line of Mary Oliver's poem "The Summer Day," I asked myself over coffee, even before breakfast, "Tell me, what is it you plan to do with this one wild and precious day?"

I had an idea, which had been incubating since Kami Ellingson took me out to see the places where restoration activities had been carried out in the Salmon River estuary, beginning in 1978. Kami, a hydrologist with the US Forest Service, had overseen a significant part of that restoration work. I especially wanted to have a close look at what she and others call the Reference Marsh, an area of tidal salt marsh that had never been diked or tide-gated to prevent the natural tidal flows and inundations. A part of it, along the natural river levee on the south bank of the Salmon, was slightly higher and drier than the rest and had been cut for hay by the Frasers. The rest of it had been grazed by their milk cows and their horses. The Reference Marsh was as wild and unadulterated a piece of tidal salt marsh as existed here. It is, for the salt marsh ecosystem of the Cascade Head Biosphere Reserve, its "museum piece," as the Neskowin Crest Research Natural Area was called by Thornton

Munger, or its *zapovednik,* to use the Russian term for a piece of a natural ecosystem preserved for scientific study.

I pulled up an image from the Cascade Head Biosphere Reserve's 2016 "Periodic Review" report that showed, on an aerial photo, the boundaries of each of the different restoration areas and when they had been restored. Then I went on Google Earth and zoomed in on the lower Salmon River. I copied an image of the right scale, and printed that, along with the image from the "Periodic Review", to use as my maps for navigating the marsh. I checked the tides at NOAA's online site and saw that if I headed out at about noon, the tide would still be going out, and I'd be there with good low water and a better chance of crossing most of the tidal channels I might encounter.

I turned west off of US Highway 101 onto Fraser Road just after crossing the Salmon River Bridge and parked at a parking area where the Forest Service has put in a few picnic tables and intends to develop an informational display. From there, I set off across the marsh toward the distant outline of Cascade Head on a network of heavily used elk trails, skirting the edge of the alder and spruce woods to the south.

It was a shining sunny day as I wandered northwest toward the river. Before I'd gone very far, I got a "crash course" in marsh topography. Sinuous channels snaked everywhere through the marsh that looked, from eye level, level. Suddenly a slit would appear in the grasses and sedges, often only a foot wide and between two and four feet deep. At the bottom of these crevasses in the low-tide marsh, trickles of salty water ran over a bottom of black muck. Spotting these small channels required real alertness, and several times I almost fell in as I suddenly encountered one. I did fall once, not into the channel but face-first into the resilient marsh vegetation on the other side.

Ah! It was a gift: the up-close smell of a dry fall salt marsh! The succulent stems of salt-tolerant *Salicornia,* commonly called pickleweed, were turning as bright red as the vine maples in the forest, and the Pacific silverweed was only a little paler. It was foliage season, marsh-level "leaf-peeping," Salmon River style.

There were occasional larger, deeply incised channels—I'm talking six or eight feet deep—that, like the smaller channels, were not visible from very far away but were basically uncrossable, at least in my state of reluctance to do some serious mud-wading. So, I followed each one up-marsh until I could find a place narrow or shallow enough to cross and then continued on in the general direction I wanted to go. After several such detours inland, I finally came to a straight line of white PVC-plastic pipes pushed into the marsh, markers for some of the scientific monitoring of marsh restoration that is still going on. The line followed a shallow north-south ditch that marked the boundary between the Reference Marsh and the so-called Y Marsh, which reaches south up Rowdy Creek, and west toward Camp Westwind, formerly owned by the Portland YWCA (and, still operating, now managed by the Westwind Stewardship Group). The marsh may be called the Y Marsh because of the camp, or because from Cascade Head the two creeks draining it look like the shape of the letter Y—I've heard both explanations. Tidal flow to this marsh area was blocked by dikes and a tide gate, installed in the early 1960s, to create pasture for dairy cows. Tide gates allow fresh water coming from inland to flow out but prevent salty water from coming in with the high tides; they dry out the marsh, especially during the nearly rainless summer season.

The dikes and tide gate here were removed in 1987, and tidal flows returned to the 155-acre Y Marsh. Even without the line of white plastic pipe that called attention to it, the boundary between the Reference Marsh and the Y Marsh was obvious. The elevation of the Reference Marsh was significantly higher, by at least a foot, and its vegetation was different and more diverse. Where I stood, on the Reference Marsh, the grasses and rushes were mostly a tawny brown, Pacific silverweed and *Salicornia* were showing red and rusty fall colors, and asters and a yellow composite I didn't know the name of were blooming. The restored Y Marsh looked greener, but had a much less diverse and more uniform vegetation, dominated by a thick-bladed marsh grass. Even after more than thirty years, the Y Marsh still hadn't recovered either its natural topography or its biodiversity. Salt marshes apparently take a long time to heal.

Tidal marshes like these trap sediments coming downriver and also accumulate a lot of plant material, which doesn't decay easily in the often oxygen-poor, anaerobic muck of the marsh. That process allows their elevation to build up over time, forming the so-called high salt marsh of the Reference Marsh. The network of drainage channels in high salt marsh evolves a dendritic, branching complexity. Because the tide was kept out of the Y Marsh for almost thirty years, less sediment was deposited and organic material had a chance to dry out and break down, so the marsh compacted and sank, and the little channels filled in.

One impetus for creation of the Cascade Head Scenic Research Area had been to prevent further development of the Salmon River estuary in order to protect its scenic values. The local residents and conservationists who pushed for its creation felt that dairy barns, grazing cows, trailer parks, amusement parks, boat basins, and housing developments were damaging the natural beauty and aesthetic values of the area, and the CHSRA Act gave the US Forest Service the authority and some funding to undo some of this development by acquiring private lands and restoring the natural salt marshes of the estuary.

When CHSRA was established, about 75 percent of the Salmon River salt marshes had been cut off from the tides by dikes and tide gates. The Forest Service's 1976 management plan for the area was to return the estuary "to its condition prior to the existing diking and agricultural use." It was a bit of a stretch for the Forest Service, which usually deals with . . . well, forests.

The sequenced restoration in the Salmon River estuary became a natural laboratory for research on salt marsh geomorphology and ecology, which began with studies by Dr. Robert Frenkel, a geographer at Oregon State University (OSU), and his graduate students. In 1978, only four years after CHSRA was established, the first dikes were removed from a fifty-two-acre area of former salt marsh in a big inner bend on the north side of the Salmon River. An OSU graduate student, Diane Mitchell, did her PhD dissertation on marsh recovery there and established permanent monitoring plots to document the recovery of the marsh. The area is now sometimes called the Mitchell Marsh.

The next large area to be restored was the already mentioned Y Marsh, in 1987, and the next, in 1996, was a seventy-seven-acre area on the north bank of the Salmon, just west of Highway 101. A second phase of restoration occurred from 2009 to 2011, when marshes that had been drained and filled for construction of the Tamara Quays Trailer Park and the Pixieland Amusement Park were reflooded. In a final phase from 2012 to 2016, tidal flows were restored in an abandoned boat basin and at Crowley Creek near Knight Park on the north bank of the river. From 2015 to 2016, through a partnership with Oregon Department of Transportation, a new culvert was installed under Highway 101 that reconnected Fraser Creek to its original sinuous tidal channel on the west side of the road. That highway, built on a levee of road fill, still blocks the natural tidal flow on the north side of the Salmon River, and restoring that portion of the estuary would be a major undertaking.

Frenkel and his students found that surface elevation is the main factor affecting salt marsh hydrology and vegetation. The Reference Marsh is high salt marsh, dominated by tufted hairgrass (*Deschampsia cespitosa*), Baltic rush (*Juncus balticus*), and Pacific silverweed (*Potentilla pacifica*). This would have been the original condition of the areas that were diked. The Reference Marsh is, in scientific terms, an experimental "control," or reference, against which the geomorphology, hydrology, and ecology of the diked and restored marshes can be measured and compared with their original, natural condition.

The phased stages of salt marsh restoration that took place over a thirty-eight-year period created a unique, unintentional opportunity for research on estuarine ecology, even though the impetus for the restoration was aesthetic and "scenic" more than scientific. This was especially fortuitous for fish biologists who later studied the life histories of coho and Chinook salmon in the Salmon River.

When I finally got to the river, it was slack water, the tide just barely starting to turn and creep back up the marsh channels. A good eight feet of undercut marsh mudbank was exposed above the low-tide water level, the smooth surface of the river reflecting sky and Cascade Head in the distance. I sat in the sun on the high bank, eating lunch and watching a northern harrier tip and drift, hunting over the marsh. No wonder the little brown sparrows were so skittish, dropping down into a marsh channel, below the surface vegetation, and scooting away before I got a good look at them. A big bold male belted kingfisher passed me, going downriver. As I started back up-marsh after lunch along a broad channel, a great blue heron *yawked* its annoyance at me for disturbing its lunchtime fishing.

I made my way as deliberately and directly as the serpentine mind of the marsh would allow—it seemed determined to foil my Cartesian navigation as much as possible—toward the head of Rowdy Creek and the site where the Tamara Quays Trailer Park levees and the raised trailer pads were removed in 2009.

~

On a wonderfully sunny winter day at Cascade Head, chilly and still, I resolved to try again to find what everyone around here calls the "tsunami layer." I had been told that on the south side of the Salmon River, where the current cuts into the layers of the Reference Marsh and the Y Marsh, forming a vertical bank, you could see a distinct layer in the marsh sediments that resulted from the huge tsunami that hit here about three centuries ago. I packed a banana and a couple of cookies for a snack, got kayaking gear together, and drove down to the River House, where the Sitka Center's canoe and kayaks are stored. The tide was running out fairly strongly when I launched the white kayak in the early afternoon, but it wasn't too bad paddling upriver against it, and I knew the cut bank I wanted to inspect would be uncovered. A couple of seals surfaced nearby, checking out any new aquatic creature in their territory as they usually do, but soon left for better adventures when they saw it was only me. Paddling upriver, I could see the Coast Range divide in the distance, and a triangular clear-cut peak with snow on it at the top of the watershed.

I first heard about the tsunami layer from Matt Taylor, the executive director of the Westwind camp. It came up when we were on the beach in front of the main camp, and Matt showed me what they call the "tsunami boat"—a medium-sized skiff, maybe twenty feet long, with Japanese characters on its white sides. It washed up on Westwind Beach a few years ago, another piece of the debris swept into the ocean after the magnitude 9 earthquake and resulting tsunami in northeastern Japan in 2011. It must have ridden around in the Japan Current for a few years before coming ashore at Westwind. Everyone here on the Oregon Coast is sensitive to the fact that they, like the Japanese, live on the edge of one of Earth's most active tectonic zones, where earthquakes and tsunamis are part of the natural environment. A "message in a boat" from that huge seismic event in Japan can't help but remind people of their vulnerability here on this coast. I really wanted to see the tsunami layer, just to be reminded, I guess, of the "big picture"—of deep time, of uncontrollable forces in nature, and maybe of my own vulnerability.

Because of some amazing scientific sleuthing, it turns out we know exactly when the last big tsunami happened, the one recorded in the sediments of the Salmon River salt marsh: it was approximately 9:00 p.m. on January 26, 1700. How, you ask, could we know the time and date of the event with that kind of accuracy? In the late 1980s, Brian Atwater, a geologist with the US Geological Survey, studied an area of dead but still-standing trees along a river on the coast of Washington. Everyone called it a "ghost forest." The trees apparently had been killed by sudden exposure to salt water, and tree-ring studies showed they had all died sometime in the winter of 1700. The geological explanation seemed to be that a major earthquake, about magnitude 9 on the Richter scale, had caused sudden land subsidence, dropping the level of the former forest floor a few feet, and into the tidal zone. In 1996, Japanese seismologists, following up on the research by Atwater, finally put a firm date on the tectonic event that killed the Washington trees. They found detailed old Japanese records that noted that "on the eighth day of the twelfth month of the twelfth year of the Genroku era," a massive tsunami struck the Pacific coast of Japan. Comparisons of dates on Japanese and Western calendars and calculations of the travel time of tsunami waves then pinpointed the Cascadia earthquake event along the Oregon and Washington coasts almost to the hour.

Chris Goldfinger, a seismologist at Oregon State who has studied the history of Cascadia subduction zone earthquakes and tsunamis, recently updated the estimate of the frequency of major earthquakes along this coast, the "recurrence interval." His evidence suggests that these events happen, on average, every 350 years in this section of the Oregon Coast.

I paddled against the slowly slackening current until I reached the exposed, river-cut bank of the Y Marsh. The marsh surface was now about six feet above water level, and an obvious line in the marsh sediments two feet down convinced me I'd found what I was looking for. It was a layer of bluish clay about two inches wide, with dark, almost black, marsh sediments below and lighter-colored sediments above. The layer itself was slick and clayey to the touch, but when I looked closely at it, smeared between my thumb and forefinger, the clay did have sparkly sand particles mixed in, maybe washed up this far from the beach by that giant wave.

The marsh sediments, as I've been calling them, were mostly heavily decomposed plant material, like soggy peat, and full of holes and tunnels made by crabs, worms, and other invertebrates. The color and texture of the sediments above and below the tsunami layer were different and distinctive. Below it, they were much darker—dark brown, sometimes almost black—and with lots of chunks of plant material, intact stems and stalks of old marsh vegetation. Above, they were paler—pale brown or grayish, and sometimes with a tinge of orange, as if they were rusty. They were also more consolidated and uniform above than below the tsunami layer, with less visible plant material. I wondered why, and an answer jumped into my head: I was looking at the cut bank of the Y Marsh, the area that had been converted to pasture for almost thirty years. As I'd seen on my earlier reconnaissance of the Reference Marsh, excluding salt water from the Y Marsh had dried it out and allowed the marsh sediments to break down and to oxidize. Now the color and texture difference of the sediments above the tsunami layer here made sense. The tsunami layer is itself a "reference layer" in the Y Marsh. It shows what happened to three hundred years of marsh sediments above it when the salty tide was excluded for thirty years.

As I looked at the layer, I tried to imagine what was happening here, human-wise. The Nechesne, the Salmon River people, had villages right here along the river, according to archeological evidence. Any village would have been inundated and destroyed by a massive tsunami wave pushing up the estuary for miles and washing back with a vengeance.

Kayaking downstream in lovely late afternoon light, I thought that if the next Big One comes now, at least I would be in a boat. I'd just turn around, point upstream, wait for the big wave, and surf it up to Rose Lodge, a town a few miles inland. Then I'd turn around and ride the exhilarating outwash surge, sweeping past Cascade Head and into the ocean. My white kayak might wash up on a beach in northern Japan in a few years, probably without me.

~

Perched on the Ring of Fire, the Pacific Northwest is a dynamic place. Besides earthquakes and tsunamis, it has and has had ice sheets, glaciers, glacial lake outburst floods, regular floods, droughts, fires, landslides, volcanic eruptions, lahars, El Niños, and Pacific Decadal Oscillations. This environmental dynamism may be responsible for the evolutionary radiation of Pacific salmon. "The salmon and the topography of western North America appear to have evolved together in response to tectonic forces that drove mountain building along the West Coast," explained David Montgomery in his 2003 book, *King of Fish.*

Salmon is the common name for a number of species of fish of the family Salmonidae; all are native to the northern hemisphere. Six species (eight if you count sea-run rainbow and cutthroat trout), all members of the genus *Oncorhynchus,* are found around the colder shores of the North Pacific, from Japan to California. Only one species, the Atlantic salmon (*Salmo salar*), occurs in the geologically much quieter North Atlantic. Understanding salmonid evolution is hampered by the fact that there are almost no fossils of these fish, but genetic analysis of their DNA suggests that the evolutionary tree of Atlantic and Pacific salmon branched about twenty million years ago when the Arctic Ocean froze. For unknown reasons, Atlantic salmon never speciated further, but by about six million years ago, Pacific salmonids had diversified into roughly their current suite of species.

Pacific salmon inhabit the freshwater streams and rivers of their home watersheds and the ocean offshore. Four species share the Salmon River: Chinook (*Oncorhynchus tshawytscha*), coho (*O. kisutch*), chum (*O. keta*), and steelhead (*O. mykiss*). They sort themselves within the watershed according to size, preferred spawning and feeding conditions, and competitive ability. Within populations of each species, variation in the timing of feeding in and migrating between freshwater, estuarine, and marine habitats results in a diversity of potential life-history patterns.

Salmon life histories are now known to be highly flexible and adaptive, but our understanding has grown only slowly through research and is still not complete. An important early study by Paul Reimers, "The Length of Residence of Juvenile Fall Chinook Salmon in Sixes River, Oregon," was published as a research report of the Fish Commission of Oregon in 1973, at the time when the debate over establishing CHSRA was at full boil. Reimers defined five different life-history types of juvenile Chinook based on the amounts of time they spent in fresh water or the estuary. The juvenile types were distinguishable by scale patterns, and by looking at the scales of adults returning to spawn, he could identify their juvenile life-history type. He found that juveniles who stayed upstream in fresh water until early summer, then remained for a period of improved growth in the estuary before going to the ocean, made up about 90 percent of the returning spawners. He concluded that estuarine feeding by juveniles was more important than had been realized previously, and his study alerted fish biologists to the importance of estuarine habitats to salmon at a time when many estuarine areas on the Oregon Coast had been lost to development—as they had been at Salmon River, except for the Reference Marsh.

In 1975, one year after CHSRA was established and a year before the area became a UNESCO biosphere reserve, the Oregon Department of Fish and Wildlife (ODFW) started a three-year study of salmon in the Salmon River, designed to provide baseline information for a new fish hatchery then being planned. The study documented Chinook and coho abundances, distributions, and life histories at a time when three-quarters of the salt marsh in the estuary had been diked for pasture, and before any marsh restoration had occurred. It provided a baseline reference condition for the salmon—an experimental control, essentially—that could be used decades later to evaluate the effect of marsh restoration on salmon populations and life histories. This study is another one of the lucky coincidences that has made Cascade Head and the Salmon River a unique natural laboratory for ecological research. The phased restoration of salt marshes in the Salmon River estuary created a serendipitous experimental setup that enabled new discoveries about how juvenile salmon use them and how important they are. Dr. Rebecca Flitcroft and her coauthors, in a 2016 article, concluded that marsh restoration "enabled a study of life-history re-emergence by Chinook and coho salmon populations, including documenting previously unknown estuary specific life-history strategies in these species."

By the mid-1990s it was clear that salmon were in trouble in the Salmon River. The hatchery hadn't helped, and many fish biologists were starting to think it was actually harming populations by changing the timing of the runs. In 1995, coho salmon in Oregon Coast watersheds like Salmon River were on the brink of being listed as threatened under the federal Endangered Species Act. To try to avoid that listing, the Oregon legislature funded the Oregon Coastal Salmon Restoration Initiative, designed to improve watershed conditions and rebuild coho populations.

In that context, Dr. Daniel Bottom and his colleagues began to look at how juvenile salmon were using the restored tidal marshes of the Salmon River estuary. Their studies started in 1997 and continued through 2003, and showed that—in a sort of salmon-themed *Field of Dreams* scenario—"if you restore it, they will come." The branching, winding networks of tidal channels that develop in mature salt marshes, like those in the Reference Marsh that I kept nearly falling into, provide a relatively safe nursery for juvenile salmon, with more places to hide from predators than in the open river or ocean, and a source of insect food. Even only one year after the last large section of marsh was reopened to the tides in 1996, they found both juvenile Chinook and coho salmon feeding in the marsh channels.

The research of Dr. Bottom and colleagues on how juvenile salmon use estuarine salt marshes involved clever experimental design, lots of patient fieldwork, and some sophisticated technology. Marking and tracking juvenile salmon throughout the watershed was part of it. Little fish were captured and marked upstream by fin-marking, with paint, at first, and later with passive integrated transponder (PIT) tags. They could then be recaptured with nets or detected with PIT-tag detectors after they had moved down into the estuary. To figure out how long the fish stayed in the estuary, Bottom and his colleagues looked at their otoliths, small stone-like structures made of calcium carbonate found in the inner ear canals of fish (and other vertebrates). These grow by daily deposits of concentric layers of calcium carbonate, somewhat like the annual growth rings of trees, and record chemical signatures of the water conditions experienced by the fish as they are being deposited. Otoliths provide a time clock that can tell how old a fish is and what was in the water around it on a given day. To make a long and complicated story short, being in salt water changes the ratio of the elements strontium and calcium in the otolith, and microscopic chemical analysis can determine exactly when a juvenile salmon entered the salty estuary from fresh water, and how long it spent there.

Combining the mark-and-recapture data with chemical analyses of otoliths enabled the salmon scientists to reconstruct the juvenile life histories of both Chinook and coho in the Salmon River. In both species there was significant variation in the timing of migration from fresh water to estuary to ocean, and also in the length of time spent feeding in freshwater versus estuarine habitats.

For Chinook salmon, a study by Volk, Bottom, and colleagues published in 2010 concluded that there is a "continuum of freshwater and estuarine life histories" that demonstrates "considerable phenotypic plasticity" in the population. Some Chinook sprinted downstream from the spawning areas and entered the estuary soon after emergence as fry. After a short period of "smoltification" in the estuary, during which their physiology adjusted to salt water, they headed out to sea, where they would spend two or three years before returning to spawn. Other fish spent longer in freshwater streams, moving down to the estuary as larger juveniles in the spring, summer, or autumn. The later migrants spent varying amounts of time feeding in the estuary before migrating to sea as late as the winter of their first year of life.

But the ODFW salmon survey conducted from 1975 to 1978, before the hatchery was opened, had found almost *none* of the early migrant types of Chinook—neither the ones that move downstream to estuarine marshes as fry soon after emergence nor fingerlings that feed in streams for a few weeks or months before moving down to the estuary in the spring. "We got a whole new perspective when we looked at the old hatchery baseline data," Dan Bottom told me during an intense download on his research over lunch in Corvallis one day. Adult Chinook that had survived their ocean life and were returning to spawn came from all of the juvenile life-history pathways, but fish who had spent some time feeding in salt marshes were especially important. According to Dan, around three-quarters of returning adults had spent more than a month feeding in the restored salt marshes. More than 30 percent of the spawning Chinook in 2004 and 2005 had been emergent fry or spring migrant types, suggesting that estuary-feeding fish made up a significant proportion of the adults returning in those years. The ODFW pre-hatchery salmon survey had serendipitously provided a baseline or reference point that allowed Dan and his colleagues to understand how important salt marsh restoration was to Salmon River salmon.

For coho salmon, Bottom and his fellow salmon scientists found that four juvenile life-history types contributed to the adult population that returned to spawn. One type fed and grew for a year in freshwater streams before migrating to the estuary for a brief period to adjust to salt water, then went to sea. This life-history type was what many fisheries managers thought was typical for coho, and they had based their estimates of adult returns on surveys of stream habitat capacity, ignoring the importance of estuaries. But the research in the restored marshes showed that there were three other types of coho that used estuaries extensively in the first year of life. Some juveniles entered the estuary almost immediately after emergence as fry, and fed and grew there for a year. Some entered the estuary early, but later moved back into freshwater tributaries to develop. And still others went downstream to the estuary after half a year in fresh water, but fed there for another few months before entering the ocean. When returning adults were sampled, the three types that used the estuary extensively made up 20–35 percent of the population returning to spawn.

"No one has done genetics on the different life-history types," Dan told me. Somehow, both in coho and Chinook, "they are able to express different life histories if the conditions [i.e., diversity of juvenile feeding habitats] allow it." This suggests that the genetic makeup of salmon allows for a lot of behavioral flexibility or plasticity—or, in the lingo of behavioral genetics, a low to moderate "heritability." Still, it seems likely that the different life-history types would have some genetic basis. Consider, for example, a 2015 study, "Juvenile Life-History Diversity and Population Stability of Spring Chinook Salmon in the Willamette River Basin, Oregon," by Kirk Schroeder and his colleagues, which found two dramatically different types of fish. One type, which they called "movers," took off and traveled up to 125 miles as tiny fry shortly after they emerged from their gravel nests, while others, labeled "stayers," hung around in those same natal streams for up to sixteen months. Starting in the same streams, it's hard to believe that it was only differences in environmental conditions, and not some genetic predispositions, that made some fish stay for more than a year and others leave immediately.

At Salmon River, life-history diversity in both Chinook and coho salmon that had been "lost," or not previously observed, reemerged when large areas of salt marsh habitat were restored. And, very importantly, the reemergence of this behavioral diversity was adaptive, in that the fish that fed in the restored salt marshes made up a significant proportion of adults of both species that later returned to spawn.

Put these conclusions together with my starting point for this story: the environmental dynamism, if not chaos, of the salmon ecosystem. Some years might be great years to feed and grow in the ocean, depending, maybe, on the state of El Niño or the Pacific Decadal Oscillation; if you sprint for the sea as soon as possible, you'll do well in life. Other years, if you go to sea right away, you might hit a warm blob or a pocket of anoxic water, and staying in your home stream for a few extra months would be a better bet. Maybe going down to feed and fatten on all those insects in the salt marsh would be a great idea—you'd go to sea bigger and faster, and get away from more predators. But maybe, if the winter had been dry, and the river flow was down, and the marsh water got too hot . . . or if a tsunami happened to hit . . . or if . . . With all the possible "or if" scenarios, what's a salmon to do?

Thinking about such questions, ecologists have recently borrowed an idea they heard about from their stockbrokers and 401(k) managers called the "portfolio effect." The idea is, if your investment portfolio has only one or two investments, and one (or two) of those crashes, you are in bad shape. Having more investments, and of diverse kinds, reduces the risk that any one thing not doing well will affect the value of your overall investment portfolio. In an article on applying the portfolio concept in ecology and evolution published in *Frontiers in Ecology and the Environment* in 2015, Daniel Schindler and his coauthors say, "Biological systems have similarities to efficient financial portfolios; the emergent properties of aggregate systems are often less volatile than their components. These portfolio effects derive from statistical averaging across the dynamics of system components, which often correlate weakly or negatively with each other through time and space."

The portfolio of life-history "investments" made by salmon spreads risks in time and space and increases their evolutionary resilience in the unpredictable corner of the biosphere they inhabit. We're throwing another risk into the mix: climate change caused by our addiction to fossil fuels. With it will probably come drier summers, more fires, warmer offshore ocean temperatures, ocean acidification, and sea-level rise that may drown the salt marshes in estuaries. If we give them a chance, salmon will hedge their bets, and survive everything that our species and the old Earth can throw at them. If we give them a chance.

6 *Art and Ecology at the Otis Cafe*

Every poet has trembled on the verge of science.
—Henry David Thoreau, journal, July 18, 1852

I met Nora Sherwood and her husband Gary for breakfast at the Otis Cafe, a famous local breakfast hotspot a ten-minute drive from the Sitka Center. One of their signature dishes is "German potatoes"—coarse-cut hash browns smothered in onions and local white cheddar cheese. They serve quarter, half, and full orders. Everyone says if you order more than a quarter order you are going to be taking some home. Their cinnamon rolls are "as big as your head," Gary said. An aficionado of Americana diners since my graduate school days, I could tell immediately that the Otis Cafe would be a place I would visit many times.

I'd first met Nora a couple of weeks earlier, soon after I arrived at the Sitka Center in the fall of 2018, one October Saturday when she was teaching a workshop on how to draw mushrooms. When I dropped into the big, barnlike Boyden Studio at Sitka, a table at the front of the room was spread with several dozen kinds of mushrooms, all picked nearby. Nora trained as a biological illustrator, and her work is precise and accurate, but artful. She has done many mushroom portraits, but always of plucked ones. I had started a conversation with her then about how artists could show ecological relationships and not just biological "objects" like mushrooms, flowers, or birds, and I wanted to continue talking about the relationship between art and ecology.

I was thinking about that idea with respect to mushrooms in particular, because the previous weekend I'd gone on a mushroom hunt with Matt Taylor at Westwind, across the Salmon River from the Sitka Center. Boletus and matsutake were exploding out of the sandy ground under the spruces and shore pines, and Matt asked me to imagine that we could see underground. The mushroom poking through the needle duff would be attached to a web of fungal hyphae, the nutrient-hungry threads that form the mycelium of the fungus, a huge net symbiosing with the roots of the trees around them. The mushroom tops we intended to eat were only a tiny part—although a beautiful and delicious one—of the whole organism. How could a visual artist capture that whole ecological portrait?

A few days after her workshop, I sent Nora a link to John James Audubon's portrait of the black oystercatcher in *Birds of America* as one example of what I had in mind. Plate 427 shows a pair of the birds on a rocky intertidal shore, one posing strong and tall on an exposed rock, and his mate crouching to jab at a limpet attached to the base of the rock her partner stands on. Audubon's naturalist's eye had noticed that the black oystercatcher was a limpet predator, and he had shown us an ecological food chain in its portrait. In fact, a fairly high percentage of the bird portraits in *Birds of America* show what they eat, or something else about their habitats; it was that ability to bring birds alive that made Audubon stand out from the other stuffed-specimen bird illustrators of his day. His great white heron holds a fish in its deadly yellow beak; the green heron is about to grab a green moth out of the air; the swallow-tailed kite clutches a small snake in its talons; a pair of red-headed woodpeckers feeds their chicks caterpillars and nuts; the golden eagle soars over stark snowy mountains carrying a limp white snowshoe hare; and our friend the bald eagle is about to rip open the white belly of a fat carp at the edge of a cascade. One of my favorites shows a house wren parent feeding a spider to its chicks, snuggled in a nest built in a holey felt hat. Audubon tells us not only something about the house wren's diet, but also that they are highly tolerant of humans.

~

They sounded annoyed, as usual. A pair of black oystercatchers, breakfasting in their home territory on another summer morning at Middle Cove, Cape

Arago, Oregon. The fog was still fairly thick as I reached the water, rising and falling gently as if it were breathing with the rhythm of the surf on the outer rocks that sheltered the cove. I heard the birds before I saw them, but scanning the mussel- and algae-covered rocks in the direction of those fog-piercing *wheep, wheep, wheep* calls, their outrageous orange bills cut through the gray morning. Sorry to disturb you again, I thought.

But I guess they were getting used to it. I was here at every morning low tide for three summers in the late 1970s, doing research for my PhD on the ecology, behavior, and evolution of the black turban snail, a common intertidal snail of the North American Pacific Coast, whose range from central Baja California to southeastern Alaska overlaps that of the black oystercatcher almost exactly.

As I picked my way over the slippery rocks to my research site, I was surprised to see someone else in the cove at this early hour. As I got closer, I recognized Peter Frank, an ecologist who taught at the University of Oregon's marine station in nearby Charleston, on Coos Bay. Binoculars and waterproof notebook in hand, Peter struck me as a naturalist-ecologist of an older era. His scientific modus operandi was patient, painstaking field observation, from which hypotheses about the causes of the observed patterns could emerge. He had been the first to notice the annual growth rings on the shells of the black turban snails I was studying, which indicated that they are very long-lived—a snail an inch across could be twenty-five years old.

Peter had been watching oystercatchers in Middle Cove for years, trying to understand the cause of a pattern that a few ecologists of the Pacific intertidal zone had already noticed. The ribbed limpet, *Lottia digitalis*—as common as my thesis snail and with almost the same intertidal range—has a striking shell-color dimorphism. Some limpets have snowy white shells, and others brown shells, but there are almost no intermediate shell colors. What could explain this dichotomous color pattern?

The best hypothesis was that a visually hunting predator was responsible, and—you guessed it—*Haematopus bachmani,* the black oystercatcher, was the prime suspect. A few years earlier a study in British Columbia had shown that limpets made up 40 percent of the summer diet of black oystercatchers there, and Peter Frank's painstaking work, published in the journal *Ecology* in 1982, confirmed that in Middle Cove, oystercatchers were also a fierce predator of limpets. Peter found that after a flock of oystercatchers had passed by on a single low tide, the "anvil rocks" where the birds brought limpets to eat them might contain forty or more limpet shells. With that level of predation, oystercatchers were clearly a significant selective agent for the ribbed limpet.

The striking shell-color differences of the ribbed limpet are related to the colors of their intertidal environment. The white-shelled limpets are found attached to the white plates of goose barnacles, while the brown-shelled limpets rest on the dark rocks around the colonies of barnacles and mussels. Both color forms are camouflaged on their respective resting backgrounds. The obvious evolutionary-ecological explanation for this camouflage dimorphism was that oystercatchers were eating, and thus selecting against, limpets whose shell color didn't match their background.

A decade after that foggy morning in the tide pools, I returned to Cape Arago to conduct my own limpet study, and again for several weeks I disturbed the Middle Cove oystercatchers—probably descendants of the ones I saw that morning when I met Peter in the cove. My interest went beyond the association of shell color and substrate color. I wondered whether the limpets knew, or cared, where they rested when they were not feeding. It seemed logical that they would "know" and "care," if evolution had anything to do with it. So, I asked them.

How do you ask a limpet where it wants to be? Simple: mark it, move it to the "opposite" habitat, wait, and watch what happens. After a few low-tide cycles in Middle Cove, the answer was unambiguous. Virtually all the white-shelled limpets that I had moved to brown rocks crawled back onto the white goose barnacles, and the brown-shelled limpets I moved onto white barnacles crawled back onto the brown rocks. I imagine them feeling irritated at the curious naturalist who disrupted their lives to ask them such an obvious question—not to mention risking their lives with the oystercatchers!

Looking back, I see now that what I was really interested in was an answer to the question, "Is behavior ecologically adaptive?" My interest in that answer mainly had to do with my own species, whose behavior a lot of the time seemed to be anything but ecologically wise. I must have thought that by asking simpler creatures like snails and limpets that question, I would become convinced that behavior *can* be adaptive, at least sometimes, before I tackled the question with *Homo sapiens.*

~

Even when I was collecting ecological data every morning those many years ago, the afternoons would often find me taking photographs or writing poems in a nearby cove that I came to call Poetry Cove. I've been puzzling about the relationship between art and ecology over all the years since those mornings in the tide pools. Although it felt natural to me to mix scientific and artistic ways of perceiving the world, when I was an undergraduate and in graduate school there was much talk of "the two cultures": science was supposedly one, the arts and humanities the other. British chemist and novelist C. P. Snow had made the phrase famous through a lecture and subsequent book, *The Two Cultures and the Scientific Revolution,* both from 1959, in which he lamented what he saw as a fundamental split in Western intellectual life between science and the arts and humanities, which he said was hindering solutions to urgent problems. Many people then thought of science and art as very different ways of experiencing and knowing the world—and many still do, it seems. They imagine that there is a gulf between science and the arts; some people even think of them somehow as "opposites." I have tried hard to understand this purported cultural schizophrenia, which I have never felt.

One first step in exploring the relationship between ecology and art could be to list some fundamental ecological ideas or principles and ask whether any art reflects those. This approach might be called the "ecology-and-art" hypothesis: *if* there is a relationship between ecology and art, *then* some art will reflect ecological principles. To start testing the hypothesis, here are a handful of some of the most fundamental ecological ideas:

- Energy flows from the sun through plants to animals.
- Materials like water and carbon cycle in ecosystems.
- Species are linked in food chains and food webs.
- Ecosystems have emergent properties because of their complex interdependencies—everything is connected to everything else, and the whole is greater than the sum of the parts.

- Humans are part of ecosystems, and now our unique behavior and technology are changing them in ways no other species ever has.

Is there any art that reflects any of these? Audubon's portraits of the black oystercatcher and many other species, illustrating ecological food chains, are good examples, of course—some initial support for the ecology-and-art hypothesis.

And what about in the literary arts? Robinson Jeffers, a now almost forgotten American poet of the 1920s to 1940s, was a major influence and inspiration for his neighbors, the marine ecologist Edward F. Ricketts and the writer John Steinbeck. Ricketts, the author of *Between Pacific Tides,* the ecologically oriented field guide to the Pacific intertidal zone I'd read before going to Camp Arago in high school, was a polymath scientist who also loved music and poetry. He was a close friend of Steinbeck, and they traveled together on a collecting expedition around Baja California and the Gulf of California in 1940, a trip memorialized in *Log from the Sea of Cortez.* Steinbeck cast Ricketts as "Doc" in his novel *Cannery Row.* Both men were fascinated by questions of evolutionary ecology, and they loved Jeffers's poetry. In his 1940 poem "The Bloody Sire," Jeffers wrote the best poetic description of evolution by natural selection that I know of:

> What but the wolf's tooth whittled so fine
> The fleet limbs of the antelope?
> What but fear winged the birds, and hunger
> Jewelled with such eyes the great goshawk's head?

Taking his lines as our lead, we could ask,

> What but the oystercatcher's sharp eye painted
> the camouflage of the ribbed limpet?
> What but the deadly orange bill destined the limpet's choice:
> the white plates of goose barnacles or the bare brown rocks?

Audubon, decades before it was formalized in scientific terms by Darwin and Wallace as the theory of evolution by natural selection, would have understood Jeffers's thrust perfectly, I suspect. In fact, I wonder if, painting oystercatchers hunting limpets, Audubon also might have guessed at the effect those birds could have on the colors and patterns of their limpet prey?

~

Audubon's ecological art served as a platform for avian conservation. He delighted in the diversity of birds, and *Birds of America* celebrated that diversity, unique to our continent. In his day, the growing global trade in feathers, plumes, and bird skins used to decorate women's hats threatened the herons and egrets he loved and painted so exquisitely. Audubon had spent more than a decade in London and Edinburgh, from 1826 to 1838, working with the best engravers in the world to produce the 435 plates for his masterpiece. London was the center of the feather trade, and more than a million heron and egret skins were sold there in one auction alone in the 1890s. The Royal Society for the Protection of Birds was founded in London in 1891 to try to stop the millinery massacre, and in the United States, the Massachusetts Audubon Society was founded in 1896. Within two years, Audubon Society chapters had been established in sixteen more states. These groups formed a loose national organization to lobby for bird protection, and by 1903 the first national wildlife refuge, Pelican Island, was established on the east coast of Florida by President Theodore Roosevelt.

So, besides illustrating ecological principles, art can inspire and motivate ecological conservation. It doesn't seem too far-fetched to hypothesize that Audubon's avian art helped to save some of his subjects from extinction. And other American artists, especially the painters of the Hudson River School of American landscape painting, are in part responsible for saving whole ecological landscapes from extinction.

Some of the works of Hudson River School painters, such as its founder, Thomas Cole, a contemporary of Audubon, tried to motivate a more harmonious relationship between humans and nature by portraying nature's grandeur and sublimity. Cole was an essayist as well as a painter, and his "Essay on American Scenery," published in *American Monthly* magazine in 1836, is a manifesto for why landscape art matters. He had this to say to his fellow Americans of the day:

> In this age, when a meager utilitarianism seems ready to absorb every feeling and sentiment, and what is sometimes called improvement in its march makes us fear that the bright and tender flowers of the imagination shall all be crushed beneath its iron tramp, it would be well to cultivate the oasis that yet remains to us, and thus preserve the germs of a future and a purer system. The pleasures of the imagination, among which the love of scenery holds a conspicuous place, will alone temper the harshness of such a state; and, like the atmosphere that softens the most rugged forms of the landscape, cast a veil of tender beauty over the asperities of life.

Cole argued that America's wilderness was one of its most distinctive features, and was worthy of preservation,

> because in civilized Europe the primitive features of scenery have long since been destroyed or modified. . . . And to this cultivated state our western world is fast approaching; but nature is still predominant, and there are those who regret that with the improvements of cultivation the sublimity of the wilderness should pass away; for those scenes of solitude from which the hand of nature has never been lifted, affect the mind with a more deep toned emotion than aught which the hand of man has touched. Amid them the consequent associations are of God the creator—they are his undefiled works, and the mind is cast into the contemplation of eternal things.

Thomas Cole inspired a generation of landscape painters, some of whom—notably Albert Bierstadt (1830–1902) and Thomas Moran (1837–1926)—traveled the western United States with government exploring expeditions and portrayed iconic landscapes like the Rocky Mountains, Yellowstone, and Yosemite in their works. Those paintings inspired eastern political and economic elites, many of whom had never, and would never, see those places themselves, to protect and preserve them. Rocky Mountain National Park, painted extensively by Bierstadt, and with one of its peaks named after him, is now a UNESCO biosphere reserve, as is Yellowstone–Grand Teton, an area painted by Moran. Both artists also painted in Yosemite, now a national park but not a biosphere reserve. The fact that Cascade Head was designated a Scenic Research Area, in part to protect its spectacular scenery, seems to fit right in to the American tradition of valuing nature in part for its aesthetic and spiritual qualities.

In addition to celebrating the beauty of nature to motivate its protection, Cole used his art to protest the destruction and desecration of natural landscapes. One of his most famous paintings, *View from Mount Holyoke, Northampton, Massachusetts, after a Thunderstorm—The Oxbow,* was made in 1836, as Audubon was about to complete *Birds of America.* From a vista looking out over the Connecticut River, Cole's allegorical painting presents a stark contrast between wild and domesticated, forests and fields, the old American wilderness and the rapidly expanding domination of it by economic and political forces. The large painting, about four feet tall and six feet wide, shows a sinuous oxbow bend in the river. On the far side, the forest has been mostly cleared for farms, and clear-cuts and wisps of smoke on the hills show that the process is continuing. On the near side, intact forest still covers the ridge, with dark rain falling in the background. Cole painted a tiny self-portrait into this painting, in the bottom center foreground, sitting at his easel but facing toward us and, apparently, the wilderness.

~

Frederic Edwin Church (1826–1900), another Hudson River School painter, was a student and protégé of Thomas Cole who eventually became more famous than his teacher. Church traveled to Ecuador in 1853 and 1857, attempting to follow the footsteps of Alexander von Humboldt (1767–1859), the great German geographer-scientist, whose ideas about complex, interdependent natural systems laid the foundation for the science of ecology. Church synthesized dozens of pencil and oil sketches from those trips to Ecuador into a giant oil-on-canvas landscape painting, *The Heart of the Andes,* which now hangs in the Metropolitan Museum of Art in New York City. The grand painting was unveiled in New York in the spring of 1859—the same year that Charles Darwin published *On the Origin of Species.* It was displayed in a huge frame that suggested a window, and Church urged viewers to stand back and use opera glasses to examine the minute details he had painted, as if they were with him in Ecuador seeing the whole landscape.

I once visited the Metropolitan Museum to see Church's grand painting. After sitting on a bench below the painting for a few minutes, taking in the landscape view—I had missed the memo to bring my opera glasses—I wandered close to the canvas to study Church's technique. It was like walking along a trail into the scene: the closer I got, the more details appeared, invisible from a distance. I was suddenly in the midst of blooming bromeliads, bright birds perched on branches, and butterflies fluttering over flowers. In the lower left of the painting, the date 1859 and the artist's name is carved on the trunk of a giant tree struck by a shaft of sunlight. A quetzal perches on a dead limb, its cascading tail juxtaposed with a small waterfall in the background.

Church hoped to take this painting to Berlin to show to his muse, but Humboldt died, in May of 1859, before he could do so. The painting was exhibited in London in September 1859, however, where it met with similar popularity as in New York. Church eventually sold it for $10,000—at that time the highest price ever paid for a work by a living American artist.

Humboldt's writings about his travels in South America were very popular and widely translated and read. They inspired not only the Hudson River School painters like Church, but also writers like Emerson, Thoreau, John Muir, and John Burroughs. Muir, in fact, was heading for South America in 1867 on his own quest to follow Humboldt's footsteps when he came down with malaria in Florida. After recovering, fate sent Muir to California; only in 1911, at the age of seventy-three, did he set off again toward South America, still after Humboldt.

Humboldt and his traveling companions almost managed to climb the volcano Chimborazo in 1801, reaching an altitude of nearly 20,000 feet, only about a thousand feet below the summit. He illustrated the distribution of vegetation on Chimborazo in a figure in his 1805 book, *Essay on the Geography of Plants*, which was far ahead of its time in combining scientific communication with creative, artistic design. Humboldt was attempting to show how the physical factors of temperature and precipitation, which change with altitude, create an altitudinal zonation of plant communities on mountains. Church's *Heart of the Andes* depicts the full range of climates and ecosystems that Humboldt had described in his 1805 "infographic."

In his masterwork, *Cosmos: A Sketch of a Physical Description of the Universe*, Humboldt included a chapter on the influence of landscape painting on the study of the natural world and expressed his opinion that it ranked as one of the highest expressions of the love of nature. He challenged artists to portray the unity of the interconnected systems that underlie any landscape—in essence to portray the ecosystem as well as its component parts. Art historian Chunglin Kwa of the University of Amsterdam, in a 2005 article titled "Alexander von Humboldt's Invention of the Natural Landscape," wrote that "Humboldt looked at plant vegetation with a painterly gaze. Artists, according to him, could suggest in their work that an abstract unity lay hidden underneath observable phenomena." Frederic Church did that.

In her thought-provoking analysis of Humboldt's influence on modern thought, *The Passage to Cosmos: Alexander von Humboldt and the Shaping of America*, Laura Dassow Walls writes that "Recovering his [Alexander von Humboldt's] cosmopolitan and multidisciplinary prospect means . . . revisioning science as an intrinsic constituent of the humanities, reading beyond the 'two cultures' to grasp a worldview that knew how to distinguish the natural and social sciences, the arts, and the humanities, but knew also how fully each interpenetrated all the others." Perhaps C. P. Snow was trying to resurrect Humboldt's holistic vision, a century later, in his critique of the "two cultures."

~

Art, according to the *Oxford Dictionary*, is "the expression or application of human creative skill and imagination." Ecology is the study of the processes, interactions, and relationships of organisms and their physical environment. In their research, ecologists certainly apply creative skill and imagination. So is ecology a kind of art? And, of course, art, like anything humans (or members of any other species) do, is one kind of "interaction" between us and our environment. The logical conclusion from these definitions is inescapable: art is one aspect of human ecology, and ecological science is a kind of art. And since art is a kind of behavior, it may be an example of ecologically adaptive behavior, the thing I was looking for in my studies of black turban snails and ribbed limpets. If the art of Audubon and the Hudson River School painters has helped save parts of our biosphere, I'd say they are contributing to human survival—and that's, by definition, adaptive.

In fact, scientists and artists are very similar in their modes of perception and methods of working. If you compare the Myers-Briggs personality profiles of artists and scientists with those of the full range of other human occupations, you will see that scientists and artists mostly fall in the same quadrant of human "personality space." The commonalities are that artists and scientists observe carefully but seek underlying patterns below the surface of sensory information, which they then abstract and represent symbolically, whether in hypotheses, paintings, or poems. For an artist or an ecologist, the underlying pattern is where the meaning lies.

Observing and searching for patterns . . . hmmm. Isn't that what even oystercatchers do, hunting for limpets on intertidal rocks? Observation and attention to pattern are fundamental to the survival of all animals. They are common to all humans, not only scientists and artists. Because evolution has tuned all of us to observe and pay attention to patterns in our environment, science and art are part of our deep evolutionary heritage, inextricably intertwined in our genes. The supposed gap between scientific and artistic ways of perceiving the world is a fundamental misunderstanding.

~

A few days after our breakfast at the Otis Cafe, Nora sent me an illustration she had just completed as part of an artist-scientist collaboration organized through the Sitka Center. She worked with a Portland Audubon Society project that was monitoring the nesting success of several species of at-risk shorebirds, including *Haematopus bachmani*, the black oystercatcher. Annual surveys of black oystercatchers have been conducted along the Oregon Coast since 2005, and the 2007 *Black Oystercatcher* (Haematopus bachmani) *Conservation Action Plan* describes the black oystercatcher as "a keystone species along the North Pacific shoreline and . . . a particularly sensitive indicator of the overall health of the rocky intertidal community."

In Nora's drawing, a parent oystercatcher was pecking at an opened mussel, teaching its fuzzy gray chick what to eat. She showed not only a food chain, à la Audubon, but also another fascinating kind of ecological relationship found in some animals: parents teaching their offspring. It's such an intimate, tender portrait, it melts your heart.

The drawing reminded me of a brilliant old study of the evolution of culture in nonhuman animals that I learned about in graduate school. Culture—the transmission of behavior from parent to offspring (or sometimes through other social channels, such as from peer to peer) through learning—is not the sole purview of our own hyper-cultural species. That study was "The Feeding Behaviour of the Oystercatcher (*Haematopus ostralegus*)," a PhD thesis presented to the University of Oxford by M. Norton-Griffiths in December 1968. It was one of the first in-depth field studies to explore the topic of behavioral genetics and sociobiology. In a nutshell—or a mussel shell, if you will—Norton-Griffiths found that the European oystercatcher exploits diverse intertidal environments by teaching its offspring what's good to eat there. He described oystercatcher "feeding traditions"—culturally transmitted knowledge about how to exploit local food chains.

Norton-Griffiths found that oystercatchers in an area with extensive beds of mussels had two feeding techniques. Some he called "hammerers." Like intertidal woodpeckers, they whacked at the mussel's shells with their beaks until the shell cracked. Other oystercatchers were "stabbers," who waited until the tide just covered the mussels, which then opened their shells so they could filter their planktonic food from the water. The stabbers took advantage of this natural response of the mussels, and with a well-aimed stab of their deadly beak through the gaped shells, they severed the adductor muscles that closed them, allowing the birds to eat the mussel with no need to break the hard shells. Through patient observation, Norton-Griffiths found that these two very different feeding behaviors were passed down in families, by parents teaching their chicks one technique or the other. This fit the definition of "culture," Norton-Griffiths argued. And the interesting thing—from an evolutionary point of view—was that the ecological flexibility provided by parent-offspring teaching and learning allowed more oystercatchers to exploit the same area of mussel beds than a population of pure "hammerers" or pure "stabbers" could have. So, the hot new idea that was almost buried in the 269 typescript pages of Norton-Griffiths's thesis was that *culture* could be an adaptive ecological strategy in evolution—and not only for humans.

And the ecological and evolutionary fascination doesn't stop there. If cultural transmission of feeding traditions is adaptive, maybe that would create an evolutionary incentive to make the transmission of parent-to-offspring learning easier. The life histories of both European and black oystercatchers seem to hint at this. In both species, parental care is prolonged far beyond that seen in other shorebirds. Oystercatcher chicks stay near the nest for five weeks or more, where parents take turns guarding them and going to find food to bring to them. When they are older, chicks follow their parents to feeding areas and are fed by them there for some time after they can fly. According to Norton-Griffiths, the explanation for this extended period of parental care "appears to lie in the specialized and difficult feeding habits of the species, which are acquired over long periods of practice by the fledglings."

Diane Bilderback, one of the Portland Audubon Society nest monitors, has been watching black oystercatcher nests in Oregon for many years, and she sent me some firsthand descriptions:

> We have often watched a parent bring a small mussel and crack it open and then pick little pieces out of the shell to drop in front of the chick. This is where the "learning" comes in—demonstrating over and over again in front of the chicks will help them "learn" how to use their beaks to open the shells. As the chick grows, it will race towards the parent as it lands so to be first if there is more than one chick. We have seen parents who are feeding three chicks specifically take food to the smaller chick to make sure that chick is getting food.

According to Diane's careful records, black oystercatcher chicks fly for the first time at between thirty-eight and forty-two days after hatching—that's up to six weeks, a very long time for shorebirds. She said that parental teaching often goes on after the chicks can fly. At low tide, parents sometimes take the chicks to an area of sand next to exposed rocks covered with mussels, barnacles, limpets, and other invertebrates; the parent pecks something to eat off the rock, while the chick watches, and feeds whatever it picked off to the chick. The chick seems to get the idea, and starts pecking at things itself. Once, during a year when mole crabs were abundant, she said, they watched black oystercatcher parents teaching chicks how to find and catch these quick, burrowing crabs on a sandy beach—thus learning how to exploit a rich but ephemeral food source.

Humans, of course, have the longest dependence of offspring on parents of any species ever, and therefore the longest opportunity for parent-offspring transmission of learned behaviors—of culture. But, with that in mind, we can almost imagine oystercatchers on an evolutionary trajectory to try to catch up with us.

~

When I was studying habitat choice and camouflage in limpets, my daughter, Anya, was five years old. Morning after morning she lay on the rock beside me, nose into a cluster of goose barnacles. She moved the white limpets to the surrounding brown rocks, and the brown limpets from the rocks to the white barnacles, after I'd glued tiny numbered tags on their shells, then helped me find the tagged ones at the next morning low tide. My son, Jonathan, one year old, splashed happily in the shallowest tide pools nearby. Beak to beak, nose to nose, parents teaching offspring. Now Anya and Jonathan are both good scientists and naturalists too. Nose to nose, beak to beak, that's the way it happens.

7 *Voices of the Old Forest*

The clearest way into the Universe is through a forest wilderness.
—John Muir, journal entry, July 1890

Thinking back to that morning in December 2018, the phrase that kept popping into my head was "listen to the voices of mom's children." What Paul Engelmeyer, manager of Portland Audubon's Ten Mile Creek Sanctuary, had actually said was that we need to "listen to the voices of the old-growth guild." But in the next sentence, he mentioned that the US Forest Service used to designate old-growth forests as Mature and Over Mature—MOM forests, for short. My mind immediately blended the two ideas, and what stuck was "listen to the voices of MOM's children."

The MOM forests were where the highest volume of wood could be cut from the smallest area, because the trees had been storing up biomass for centuries. MOM forests were old forests, like the one on Neskowin Crest that Fremont McComb had called "decrepit old forests" in his report to Thornton Munger in 1941. The "OM" part of the designation is a giveaway for the view of the value of forests through the lens of the wood-production-oriented forest managers at the time. "Over mature?" It would be like a public health official saying, "OK guys, these retirees and senior citizens, they aren't contributing anything, obviously. Can't we just take them out of the retirement homes, out of the extended care facilities, pull them off the oxygen and dialysis, and let them die already? Make way for the young ones that are growing fast and are hungry for jobs?" The industrial foresters knew the dollar-to-effort value of cutting the old growth and targeted it as their highest priority. It was a conspiracy, really, from the 1950s to the 1980s, between the US Forest Service and timber industry, to take as much value from the old forests left on public land as possible, as quickly as possible.

We stood on the bank of Ten Mile Creek and looked across to the south, up a steep slope with giant trees, several with broken tops where side branches had taken over as leaders and grown new vertical trunks. Paul said you can see marbled murrelets plunging in to their nests right here on a summer dawn, silhouetted against the sky, bringing food to the chicks. Giant bigleaf maples stood on the floodplain behind us, draped like druids with robes of moss. In the rushing riffle in front of us there was a fresh redd—a nest circle of clean silver gravel constructed by spawning salmon in the otherwise algae-covered brown streambed.

Ecologists use the term "guild" to refer to a group of species that all make their living in a similar habitat, or in more technical terms, a group of species that "overlap significantly in their niche requirements without regard to taxonomic position." The word is borrowed from the term for the medieval associations of craftsmen who controlled traditional crafts such as weaving, glassblowing, metalworking, woodcarving, and so on. Paul listed some other members of the old-growth guild: joining the marbled murrelet are the northern spotted owl, red tree vole, northern flying squirrel, and Humboldt marten. Each of the species in the old-growth guild is an indicator species for old-forest health and intactness—each a different color of canary in the coal mine, maybe we could say. Together, bonded in solidarity by their requirements for an old-forest niche, they form an indicator guild. Theirs were the "voices" Paul said we needed to listen to. "Ask each of them, what do you need? What kind of habitat do you need?"

Since none of those species can speak human languages, scientists and conservationists have to translate for them and be their voices. Paul has been an interlocutor for the old-forest guild for more than forty years. He got into forestry work through the Hoedads Reforestation Cooperative, a worker-owned tree-planting cooperative that started in 1971 in Eugene, Oregon. He learned to climb big trees to harvest cones for tree seed and later put that skill to work in a study of marbled murrelets, with the team that in 1990 took the first video of an adult murrelet brooding and feeding a newly hatched chick.

So how are members of the old-growth guild doing? One credible source for the latest information on their status is "PNW-GTR-966"—General Technical Report 966 of the US Forest Service's Pacific Northwest Research Station, titled *Synthesis of Science to Inform Land Management within the Northwest Forest Plan Area,* a three-volume report of more than a thousand pages published in 2018—which reviews the current state of the science related to forest management in the Pacific Northwest.

Let's start with the marbled murrelet, *Brachyramphus marmoratus.* Marbled murrelets belong to the seabird family that includes puffins, auks, murres, and guillemots, but they are unique in nesting in old coastal forests up to about fifty miles inland—which is what makes a seabird an unlikely member of the old-forest guild. Small robin-sized birds, they use their stubby wings as paddles underwater—basically "flying" in water—as they chase the small fish and krill they feed on near the coast. In the air, they fly very fast—around sixty miles an hour, probably just to keep their chunky bodies airborne on such short wings—as they commute from ocean to forest and back while feeding their chicks. Logging of old growth reduced their populations dramatically, and they were listed as threatened under the Endangered Species Act in 1992. Although the Northwest Forest Plan (NWFP) of 1994 focused more on the northern spotted owl, recovery of the marbled murrelet was also an important objective.

Monitoring of murrelets between 2000 and 2015 in the entire NWFP area, from San Francisco to the San Juan Islands, gave a total population estimate of around 21,000 birds and showed a slight downward trend, according to the *Synthesis of Science.* More than one-third of that population is found along the Oregon Coast, between Coos Bay and the mouth of the Columbia River, and does not seem to be declining in that area. Surveys of murrelets in the nearshore ocean found the highest densities between Florence and Newport, and a Marbled Murrelet Important Bird Area was recently established by the Audubon Society just north of the Ten Mile Creek Sanctuary, centered around Cape Perpetua in the Siuslaw National Forest.

Populations of the fish and crustaceans that murrelets eat can be affected by ocean conditions such as water temperature, upwelling, and algal blooms, which in turn may be affected by climate change. Recent research has shown that murrelets have the best nesting success in large patches of old growth, with more "interior" and less "edge," and suffer when those areas are fragmented by logging or roads. Predators of murrelet eggs and chicks, such as jays and crows, can penetrate farther into smaller patches and find murrelet nests more easily. In short, although murrelets in central Oregon may be holding their own, they face many threats and unknowns. Ongoing monitoring is needed, and the goal of stabilizing and increasing murrelet populations has not yet been achieved.

What about the northern spotted owl (*Strix occidentalis caurina*), the bird whose listing as threatened under the Endangered Species Act in 1990 was central to the controversy that the Northwest Forest Plan tried to address? This was the bird featured on bumper stickers in timber country that read, "Save a Logger—Shoot an Owl."

The Northwest Forest Plan tried to balance the habitat needs of the owl with society's need for wood products. Given the dramatic imbalance that existed at the time, when most old growth had already been cut, rebalancing these needs required a major effort to protect remaining old forests and restore them where possible, and that required a dramatic change in the timber industry in the Pacific Northwest. Jerry Franklin was one of the architects of the NWFP; his research background and interest in forest succession and old forests made him an obvious choice for that role. He worked with a large interdisciplinary team, called the Forest Ecosystem Management Assessment Team, to craft the science-based plan, which was adopted by President Bill Clinton.

Because they were a focus in the big battle between two different views of the value of forests, northern spotted owls became one of the most studied birds in the world. Research showed that their preferred habitat is forests 150 to 200 years old or older. Although forests at least 125 years old are suitable, anything younger is marginal. Since 1994, when recommendations of the NWFP began to be implemented, protecting as much of the remaining old forests as possible has not stopped the owl's decline, and not enough time has elapsed to restore much old growth.

In the meantime, another significant threat has emerged: competition with a closely related species, the barred owl (*Strix varia*), native to eastern

North America. Barred owls have been expanding their range westward for the past century, until, at just about the time spotted owls were becoming endangered from loss of their old-growth forest habitat in the Pacific Northwest, they showed up there. Barred owls are ecologically similar to spotted owls, but are slightly larger, have a more varied diet, use a broader range of forest conditions for nesting and foraging, are more aggressive in defending territories, and have higher survival rates—tough competitors, in other words, for the northern spotted owl. The 2018 *Synthesis of Science* found that barred owls now occupy the entire range of the northern spotted owl, and that "barred owl densities may now be high enough across the range of the spotted owl that despite the continued management and conservation of suitable forest cover types under the NWFP, the spotted owl population will continue to decline without intervention to reduce barred owl populations."

Forest managers appear to be facing a difficult ecological and philosophical dilemma. Do we need a new bumper sticker, "Save a Spotted—Shoot a Barred"? The barred owl's arrival as a "native invader" appears to be due to a combination of human modification of ecosystems, over a century or more, and climate change. How guilty should we feel, and can we really turn the clock back now for the spotted owl?

Northern flying squirrels (*Glaucomys sabrinus*) and red tree voles (*Arborimus longicaudus*) are important prey of spotted owls. They need big old trees and a connected forest canopy, so forest fragmentation and some types of thinning reduce their populations. Fungi such as truffles are key food sources for northern flying squirrels, which disperse the truffle spores by eating the underground fruiting bodies—just one example of the web of interconnections in the old forest. The Pacific marten (*Martes caurina*), a member of the carnivorous weasel family, is part of the old-growth guild as well. It prefers large patches of late-mature and old forest, with the complex vertical and horizontal structure where some of its favored prey—flying squirrels and red tree voles—are found.

The main threat to all the members of the old-growth guild is loss of old-forest habitat, which the Northwest Forest Plan was designed to stop and reverse. At that time, according to Jerry Franklin and other forest scientists, only about 2 percent of the landscape west of the Oregon Coast Range still had old forests, and only 14 percent had mature forests, compared with their estimate of 44 percent old and 17 percent mature forest cover before Euro-Americans arrived on the scene. No wonder the voices of MOM's children were crying for help.

In order to reverse the ecological effects of that dramatic loss of old forests as quickly as possible, the US Forest Service turned to active management of cutover lands, using silvicultural treatments like thinning, planting, and controlled burning to try to restore the complex stand structure of natural forests as rapidly as possible. When combined with other treatments such as topping trees to create lateral growth, girdling to create standing dead trees, and thinning but leaving logs on the ground to increase woody debris, this regime is sometimes referred to as "ecological forestry." The problem is, ecologists hardly know enough about the exact habitat requirements of old-forest species to know how they will respond to these supposedly science-based silvicultural treatments, so research has to go hand in hand with restoration—as it did in the restoration of salt marshes in the Salmon River estuary.

~

After a natural disturbance such as a fire, forests develop and change in an unfolding continuum—a sequence of stages of increasing complexity, species composition, three-dimensional structure, and functional processes—called ecological succession. The roots of this idea of directional ecological change reach deep into the history of American ecology. Henry David Thoreau became an early expert on ecological succession after he and a friend went on a camping trip in 1844 (he was then twenty-seven) and accidentally started a forest fire that almost burned down the town of Concord, Massachusetts. His careful observations over the next decade or so convinced him that the local pitch pine and oak forests, with their huckleberry and blueberry patches, were the result of frequent fires, and that Native Americans had deliberately used fire to shape and manage them across the landscape. He gave several lectures on the topic to farmers around Concord in 1860, and prepared an essay, "The Succession of Forest Trees," which was widely read and reprinted. Aldo Leopold and Thornton Munger were well aware of the concept a century ago.

Infrequent stand-replacing fires are the main ecological disturbance that resets forest succession in western Oregon. Fire return intervals west of the Oregon Coast Range were two hundred years or more before Euro-American colonization, and based on that statistic, almost half of the landscape would have been old growth, undisturbed for two centuries or more, and another fifth would have been mature forests, undisturbed for one hundred to two hundred years. In the area on the Neskowin Crest that Fremont McComb assessed for Thornton Munger in 1940, as it was being considered for designation as a research natural area, McComb reported that 73 percent was old-growth spruce, hemlock, or a mixture of both species, and 27 percent was mature second growth. In other words, the Neskowin Crest was 100 percent old and mature forest, truly the "museum piece" of old coastal forest that Munger had been looking for.

Dr. Thomas Spies, a protégé of Jerry Franklin who worked with him on the 1994 Northwest Forest Plan and led the 2018 *Synthesis of Science* that is being used to update the plan, wrote in 2009 that "the only places where scientists can learn about the structure and process of old forests is where the absence of human activity has allowed some populations of trees to survive for centuries—so-called *virgin* or *near-virgin* forests. At a grand scale these are the scientific *controls*—places against which to compare the changes that result from the unplanned 'experiments' of human activities." Through a series of unique historical events described in earlier chapters, the Neskowin Crest Research Natural Area, in the Cascade Head Experimental Forest, in the Cascade Head Scenic Research Area, in the UNESCO Cascade Head Biosphere Reserve, has become one of those scientific controls—a place to understand the "unplanned experiments" of our species on the biosphere.

~

"Arriving at NC-08," read the message on my Garmin GPS as it alerted me with a beep. I had typed in the easting and northing UTM coordinates of plot NC-08 from the list Rob Pabst had given me into the hand-held global positioning system unit when I started on the trail toward Hart's Cove from Neskowin. I looked around, and immediately noticed a metal tag on a tree about ten feet above the trail. Wow!—I thought. I found it: Neskowin Crest Research Natural Area Plot Number Eight.

Rob Pabst is the forest scientist at Oregon State University now in charge of monitoring all of the permanent research plots in the Cascade Head Experimental Forest and Neskowin Crest Research Natural Area. Data from these plots is a treasure trove of information for long-term ecological research on coastal temperate rain forests. The array of forty-four permanent forest monitoring plots on Neskowin Crest was the brainchild of Jerry Franklin. In 1979, back at the Forest Service's Pacific Northwest Research Station and in charge of the Cascade Head Experimental Forest after his stint at the National Science Foundation in Washington, DC, Jerry sent some technicians out to set up plots along west-to-east transects in the Research Natural Area. "They came back saying 'No way! Moving through that old forest is too rough!'" Jerry told me. "I sent them back out!" The plots were finally established and are still being revisited and remeasured every five or six years by Rob and his research crew.

At NC-08, the metal tags were affixed to every tree in sight with big galvanized nails. Each tag—silvery, but rusting at the bottom—had a number stamped into the metal. The understory was a pick-up-sticks jumble of fallen trunks of all sizes and states of decay. Bashing off through chest-high sword ferns, clambering over fallen logs, stepping on a mossy forest floor that looked solid but that kept giving way and dropping my leg hip-deep into a tangle of decaying wood and branches, I eventually found about thirty tags. This was "nurse log" territory, and I was fascinated that the more recently fallen, more solid trunks were where the baby trees were taking root. Most of the tiny trees were hemlocks, but in a few places where the fall of a big tree had created enough of a light gap, Sitka spruce were springing up too. Quite a number of small trees, six inches in diameter or less, were dead, tops broken off. So, the plot data should be showing the sprouting of young trees in the understory, the demise of most, and the survival of a few. Without another disturbance, the bet was on the hemlocks.

Down off the trail about fifty feet, perched on a 45-degree slope above a singing stream, was my favorite: no. 384, a giant Sitka spruce more than six feet in diameter. I made a note to ask Rob for the data on this old one. In 1979, when it was first measured, it was 185.8 centimeters in diameter at breast height; when last measured, in 2015, it had grown to 190.1 centimeters—only 4.3 centimeters, less than two inches, in thirty-six years. Old trees grow slow.

Rob also sent a summary spreadsheet with data for all the trees in Plot NC-08. There has been a bit of turnover in the understory since 1979, with twelve young trees coming up; eleven were western hemlock and only one a Sitka spruce. Overall, the number of trees in the plot decreased slightly, from thirty-eight to thirty-three. Everything at NC-08 suggests a pretty stable old forest.

After thrashing around just off the trail at NC-08 for almost an hour, I went on, contouring over to a stream crossing, and then out along the contour again to the next ridge, where the trail swung around it and doubled back on the contour. The map showed another permanent plot, NC-09, right there. I put the UTM coordinates into the GPS, and sure enough, right when I was expecting it, I got another "arriving at" message and beep. Here, a bunch of small trees, four to six inches in diameter, were tagged. Recruitment from the understory, all shade-loving hemlock.

I was starting to get a firsthand understanding of just how challenging it is to monitor forty-four permanent plots across this rugged landscape. The two plots I'd found were virtually on the trail, but most of the plots are not. You would have to brush-bash and moss-wallow off trail, following your GPS's directions, to find most of them. Long-term ecological research in a forest like this is obviously a strenuous undertaking. Once you reach the plots, how long would it take a crew to measure each tagged tree, I wondered? And when you get home with the data, what kind of analysis would be needed to say, "Sitka spruce dying out, western hemlock filling in," or "growth slowing in last decade," or to come to some other conclusion? Will these plots be able to tell us if climate change is altering forest growth rates or forest succession here? Is all this "science" worth it?

I posed that question to Jerry Franklin. He argued that, because forest succession takes so long and is so sensitive to factors that we may not appreciate or be measuring, we need these long-term ecological records to improve our understanding and forecasting ability. Without ground-truthing in actual forest plots, the data that go into ecological models are worthless, he said.

~

Red alder (*Alnus rubra*) is an early successional species, springing up after a forest-clearing disturbance like a fire, doing its thing for a century, and then giving up the territory to the shade of taller, longer-lived conifers. Red alders fix nitrogen, and nitrogen is essential to life. It's a key element in building proteins and the enzymes that control biochemical reactions in all organisms. Elemental nitrogen is abundant in the biosphere, making up about 78 percent of the atmosphere, but as such it is not useable by most organisms, which require chemically "fixed" nitrogen, usually in the form of nitrate or ammonia. A small number of organisms are able to capture and fix atmospheric nitrogen, a process that requires energy. Nitrogen-fixers include some non-photosynthetic bacteria and some photosynthetic cyanobacteria. Many of them form symbiotic relationships with other organisms, trading on their ability to provide a needed nutrient to their partners in return for energy, protective habitat, or other nutrients. These symbiotic swaps occur in association with the roots of some plants, one such being red alder.

Nitrogen is often a "limiting factor" in ecological systems. Because it is relatively scarce in the fixed chemical form that organisms need, it may control, or limit, the amount of life that can exist in a given place. The ability of red alder to host nitrogen-fixing symbionts around its roots may be an evolved adaptation that gives it an early competitive edge in colonizing forest areas disturbed by fires or windstorms.

Because of all this, I wanted to see the "alder-conifer plots," site of a pioneering experiment in forest ecology that began here in the Cascade Head Experimental Forest in 1935, where the natural regeneration of red alder and conifers on an abandoned potato farm was turned into a clever experiment. Alders were cleared and only conifers left on one plot; conifers were cleared and only alders left on another; and both kinds of trees were allowed to regenerate naturally on two more plots. These ALCO plots, as they are called in the Experimental Forest's acronymic code, reveal an ecological mind behind their establishment—probably Thornton Munger, the first director of the Experimental Forest.

Jerry Franklin was a keen student of forest succession, and he had long recognized the value of the alder-conifer plots. In 1968 he published a study with three coauthors titled "Chemical Soil Properties under Coastal Oregon Stands of Alder and Conifers," based on their research in the ALCO plots. They found a lot more nitrogen in the soils under alder or mixed alder and conifer stands compared with the pure conifer stands, and discovered that alder will fix significant quantities of nitrogen even on sites that already have high nitrogen levels, creating a soil-nitrogen bank for the future conifer forest.

I parked at a pull-off less than two miles up the road from the historic Cascade Head Experimental Forest headquarters. My Garmin GPS unit indicated that the plots were closer than the length of a football field, so I figured even if it meant climbing over jumbles of moss-matted fallen logs and bashing through salmonberry thickets, I could probably make it there. A faint trail led to the southeast, in the right direction. I plunged into the forest.

I found the alder-only plot first, its boundary flagged with orange plastic tape, and stood among octogenarian alders—ancients of their species, but with trunks less than a foot in diameter. Looking up to the gray sky, I could see that a significant number, maybe one in ten, were already dead. But I noticed that the trunks of those tall, fragile dead ones were the perfect environment for oyster mushrooms (*Pleurotus ostreatus*), fungal decomposers that make a living by feeding on the dead wood of alders. I picked a few handfuls, and their smell lingered on my fingers all afternoon, the aromatic essence of the place: rain and moss, fern and resin, and a whiff of salmonberry.

That night I sautéed them for dinner in a little butter. The first bite was a flood of subtle *umami*—a Japanese word that describes a savory taste, sometimes called the "fifth taste." Western scientists had decided that humans could taste only four tastes: sweet, sour, salt, and bitter, and they were skeptical when a Japanese scientist, Kikunae Ikeda, proposed in a 1909 paper that people could also taste umami. It wasn't until the 1980s that the Western scientific community began to suspect that Ikeda was right, and the discovery of taste receptors for umami using molecular genetic techniques in 2000 finally convinced them. But, of course, we all knew we could taste umami, and we had liked it even before the scientists figured it out. It is the alluring, meaty taste of amino acids—the nitrogen-containing components of every protein—especially glutamic acid. Umami is essentially the taste of protein, and protein is essentially the taste of nitrogen, that minor component of our atmosphere that builds the framework of all life. At least to my taste buds, real oysters—the marine molluscan ones—have a hyper-umami taste, so maybe it makes sense for "oyster" mushrooms to be loaded with umami too.

I ate slowly, savoring the flavor of forest succession and symbiosis.

~

The 1994 Northwest Forest Plan was an attempt to reconcile the habitat needs of the endangered old-growth guild and the need for timber production in the Pacific Northwest. The 2018 *Synthesis of Science* concluded that, although the plan protected remaining old-growth habitat, the spotted owls, marbled murrelets, coho salmon—and timber harvests—have continued to decline. New challenges, not fully foreseen in 1994, have come to the forefront, including barred owl–spotted owl competition and the loss of early successional species because of fire suppression. It also found that the policy, social, and ecological contexts have changed since 1994, and that although federal lands are essential for restoring species listed under the Endangered Species Act, actions on federal lands alone won't be enough. If what the *Synthesis of Science* calls "ecosystem goals," such as ecological resilience, are to be achieved, the balance between humans and nature must come at a larger landscape level that includes federal, state, and private lands.

We need reconciliation between humans and nature if we want resilience. But how do we get there? Much research has been done, but more is needed. Scientists, conservationists, and land managers are trying their best to speak up for the old forest and her children. Cascade Head has been an important place for listening to their voices and a laboratory for learning their languages. Let's keep learning and listening.

8 *So Long, Silverspot*

And ere a man hath power to say "Behold!"
The jaws of darkness do devour it up:
So quick bright things come to confusion.
—William Shakespeare, *A Midsummer Night's Dream*

The Nature Conservancy's Cascade Head Preserve was established in 1966, in one of the early victories to control development along the Salmon River estuary and on the scenic south-facing meadows above it. Those meadows were an example of a coastal prairie, a unique habitat along the Oregon Coast that was becoming increasingly rare. Coastal prairies exist because of "a magical mix of salt air, unusual soils, and grasslands maintained by fire and grazing," says an interpretive sign at Cannery Hill in the Nestucca Bay National Wildlife Refuge, where active restoration of coastal prairie is under way. These coastal prairies are the habitat of the Oregon silverspot butterfly, *Speyeria zerene hippolyta,* which was listed as threatened under the Endangered Species Act in 1980 (more later on the butterfly and its status today).

When I arrived at the Sitka Center at the beginning of October, a patch of the south-facing meadows on Cascade Head had been blackened by a recent fire. Seen from Westwind Beach, the edges of the burn were suspiciously straight, and I guessed it was a prescribed burn. Hiking up from the Sitka Center one afternoon a few days later confirmed my suspicion. Small-diameter cloth fire hoses that had been used to control the fire were still positioned along the trail. I checked with Debbie Pickering, TNC's Oregon Coast ecologist and the manager of the Cascade Head Preserve, and she filled me in on the details of the burn. She mentioned that she was going out in a few days to scatter native grass seeds over the area, and I volunteered to help.

On a clear mid-October morning we met in the meadow below the Sitka Center and hiked up to the burn, lugging our gear: two hand-cranked seed spreaders, a big bag of local native grass seeds Debbie and volunteers had harvested earlier, a hundred-foot tape measure, lunch, and plenty of water and sunscreen. The seeds were California brome, *Bromus carinatus,* a tall native bunchgrass, good for stabilizing bare soil with its deep roots. It sprouts quickly, and the idea was to give a native grass a competitive boost against the aggressive nonnatives. The first step was to flag out a measured area and fill the spreaders with the amount of seed needed to distribute about one seed per square foot, the optimal scatter according to grassland restoration experts. The fire had burned back clumps of native coyotebrush and of Himalayan blackberry, an invasive alien, but pulling the measuring tape or trying to walk in a straight line cranking the seed spreader was still a challenge. The blackened blackberry canes were still mean, their thorns as sharp as before the fire, catching at my old jeans as I scrabbled across the steep, charred slopes. We measured out areas and cranked out seeds, stopping for lunch with a grand view of the estuary and the coast spread out to the south. At the end of the warm afternoon, we'd seeded a little over two acres of the twenty-seven-acre burn. We pulled the last sections of hose that had been used during the burn down the trail to Debbie's car as the sun was about to set on a platter of silver ocean to the southwest. It had been a good, long day of hot, hard work. For what?

The objective of the prescribed burn was to slow the spread of woody shrubs like coyotebrush and nonnative species like Himalayan blackberry into this remnant of coastal prairie. Coyotebrush (*Baccharis pilularis*) is a woody native evergreen shrub in the sunflower family, found from Baja California to only about here; Tillamook County, Oregon, is the northern end of the species range. It has a successional talent for invading and colonizing grasslands in the absence of fire or grazing—as it was obviously doing here on Cascade Head. Himalayan blackberry, *Rubus armeniacus,* is another, complicated, ecological story. Native to Armenia and northern Iran, it was deliberately introduced in Europe in the early 1800s, and in North America in 1885, because of its large, delicious berries. It soon escaped from cultivation and became established as an invasive species—nowhere more so than in Oregon. Birds and rodents love the fruits and spread the seeds, and this species is an ecological competitor that can't really be beat—but burning can set it back a bit as it tries to overtake the salt-spray meadows of Cascade Head. The fire had worked its medicine, setting back the shrubs, burning off the thatch of dead grasses—many nonnative—and opening the ground for violets and other native forbs. The idea is that the reintroduction of fire is the prescription for restoring coastal prairie.

~

Seventeen years after TNC's Cascade Head Preserve was created, in 1983, the first thorough study of the ecology of Oregon coastal prairies was published, a PhD thesis by James Ripley from Oregon State University, titled "Description of the Plant Communities and Succession of the Oregon Coast Grasslands." Ripley undertook his extensive survey of coastal prairies because, although they were found at many places along the Oregon Coast, they had not been studied much and they were disappearing rapidly. The only previous research was a 1967 master's thesis, also from Oregon State, by Eric Davidson, who studied the then-ten-acre prairie above Hart's Cove in the Neskowin Crest Research Natural Area. In the introduction to his thesis, Ripley wrote that "beyond Davidson's study, very little else is specifically known about the Oregon coastal grasslands and only limited speculation has been proposed to explain their ecology. The absence of detailed ecological information . . . represents a gap in our knowledge." Ripley surveyed coastal prairie plant communities at twenty-four locations, from Cape Falcon in Tillamook County south to Cape Ferrelo in Curry County. He found a total of 249 native and introduced species and used sophisticated multivariate statistical analysis to characterize the vegetation. His sampling locations included seven plant transects in and around the TNC Cascade Head Preserve, including the Pinnacle and the area where Packwood peeked over the cliff and met the eagle in 1973.

Ripley found that a few of the plant associations of coastal prairies appear to be "climax" communities—that is, able to persist indefinitely without disturbances, usually because of the already stressful ecological conditions there. For example, what he called the slough sedge (*Carex obnupta*) community was found only on very wet sites, such as ravines with fine textured, poorly drained soils. A ravine below the Cascade Head Pinnacle supports this plant association. Another stable plant community found in coastal prairies is dominated by the coastal strawberry (*Fragaria chiloensis*), sea thrift (*Armeria maritima*), and several other salt-tolerant native plants. This community is often found along the edges of headlands and cliffs, where these plants can successfully compete in an environment of high wind and salt spray exposure.

The most common plant community found in areas that were formerly grazed are dominated by velvet grass (*Holcus lanatus*) and sweet vernal grass (*Anthoxanthum odoratum*), both invasive perennial grasses introduced from Europe. Both have a fairly high tolerance to salt spray, so can occur fairly close to the ocean. Ripley considered this plant community to be "an intermediate seral stage in a general successional pattern from the disclimax represented by grazed pastures to either a climax shrub or tree dominated community." In other words, without further grazing, coastal grasslands of this type would be invaded by shrubs and eventually trees. "In a number of locations, such as at Cascade Head, the advance of trees and shrubs into this community can be readily observed," he wrote.

As I did to find the exact spot of Packwood's view from the 1973 newspaper photos, Ripley used historic photos to find and rephotograph sites with coastal grasslands, documenting how they had changed over time. Especially striking to me are his repeat photos of Neahkahnie Mountain. In a 1900 photo, the entire south slope of the mountain was grassland; Ripley's 1983 photo shows that forest had by then filled in most of the slope. In photos of Cascade Head taken around 1900, its southern slopes are meadows, and it's obvious why it was once called Bald Mountain or Grass Mountain.

Neahkahnie Mountain was a sacred mountain of the Tillamook people, its name in their Salishan language said to mean "place of god." Oblivious to this history at the time, my cousin and I hiked up Neahkahnie Mountain almost every summer from Manzanita during my teenage years, when my family visited our Oregon relatives. David and I scrambled up the slopes of salal and blackberry and arrived scratched and sweaty at the top to take in the sublime view. Maybe, in retrospect, it was a youthful vision quest or pilgrimage of a sort.

A time series of aerial photos of Cascade Head, the Salmon River estuary, and the headlands north of Lincoln City are presented in a report from an initiative called the Coastal Prairie Project. The first photo is from 1952, the second from 2016. When the current boundaries of prairie and forest are overlaid on the first photo, the change is hard to believe. Large areas just north of Lincoln City that were coastal grassland are almost completely forested, as are some slopes north of the Salmon River. A rough visual estimate from the overlay suggests that in those sixty-four years, more than 80 percent of the former grasslands were lost. Only the headland meadows of the TNC Cascade Head Preserve have held their own; they appear to have lost only about 10 to 15 percent of their former area from forest encroachment around the upper edges.

Ripley had an applied interest in closing the gap in our knowledge of the ecology of Oregon's coastal prairie ecosystems. In his thesis he wrote, "It is my hope that this study will provide useful information on the nature of this attractive and useful vegetation which will suggest more intelligent methods for its present and future management." As with salt marshes, salmon, and spruce-hemlock forests, research is needed to understand the dynamics of ecosystems so that they can be conserved or restored. Research and restoration must be linked.

Oregon coastal prairies, such as those that dominate the Pinnacle and south-facing slopes of Cascade Head in the TNC Preserve seem to require fire and grazing to persist. Without those disturbances, ecological succession would take place, shrubs and trees would encroach into the meadows, and they would eventually become Sitka spruce–western hemlock forests like those that dominate the rest of this landscape. Ripley wrote that "Indian-set fires were probably the major factor responsible for the maintenance of coastal grassland vegetation prior to European settlement." He found charcoal in 33 percent of the soil samples collected from his study sites, suggesting that in many places the grasslands were created or maintained by fires in former forest or shrubland. Historian Stephen Beckham, in his 1975 report on the history of Indian and white settlement of the Cascade Head Scenic Research

Area, suggests that burning was practiced for practical ecological purposes: to maintain habitat for favored food plants and berries and to attract game and facilitate hunting. The Coastal Prairie Project report with the aerial photos called coastal prairie the First Peoples' Prairie—using a term common in Canada to refer to indigenous Native Americans. The report explained, "The arrival of European settlers was detrimental not only to the first peoples' population size through disease and conflict. It also removed their interplay with the landscape and the disturbance of fire which many coastal prairie species evolved with and are promoted by."

Because both fire and grazing were—at least in part—the result of human activities, these Oregon coastal prairies are what have been called "anthropogenic ecosystems." In that regard, they are like the "oak openings" of Wisconsin—oak savannas created and maintained by Native American burning—that so enchanted the young John Muir at his boyhood farm at Fountain Lake, and Aldo Leopold at his shack along the Wisconsin River. The huge miombo woodland ecoregion that covers parts of eleven countries in southern Africa is a fire-dependent ecosystem too—one that has probably been shaped by our ancestors' use of fire for hundreds of thousands of years. Anthropogenic ecosystems like these pose a thought-provoking philosophical question: Are they natural, or unnatural? If we agree that humans are part of nature, I think we would have to call them natural.

~

The Oregon silverspot butterfly (*Speyeria zerene hippolyta*) is a small, darkly marked coastal subspecies of the zerene fritillary, a widespread butterfly species found mostly in western North America. The zerene fritillary is one of sixteen species of the genus *Speyeria,* in the butterfly family Nymphalidae. The larvae of all species feed on violets. For the Oregon silverspot, the larval host plant is the early blue violet, *Viola adunca,* found in coastal prairies. In late summer, adults emerge from the chrysalides where they have metamorphosed from caterpillars; they fly, feed on nectar from flowers such as yarrow, goldenrod, aster—and also the nonnative invasive, tansy ragwort—and mate. Females lay eggs in leaf litter near violets, and tiny larvae hatch about twelve days later. After eating their eggshell, a high-protein snack, larvae wander off to find a place to sleep through the winter, a kind of hibernation that, in insects, is called diapause. When they become active again in the spring, silverspot larvae feed exclusively on violet leaves for two months or more until they are large enough to pupate, forming a chrysalis. The next generation of adults emerges in August and September to begin the cycle again. Dr. David McCorkle, now a professor emeritus of biology at Western Oregon University, was the first to realize that the Oregon silverspot was a coastal prairie specialist, and he initiated the Endangered Species Act listing petition for it in the 1970s.

At the time of its listing, the only known viable population of Oregon silverspots was in the Rock Creek area south of Cape Perpetua in the Siuslaw National Forest. Historically, it was found along the Washington and Oregon coasts from Grays Harbor south to about Heceta Head, and as a disjunct cluster of populations north of Crescent City, California. At least twenty separate localities were known for the butterfly in the past, when its coastal prairie habitat was much more common. The subspecies was first found at TNC's Cascade Head Preserve in 1988, although it undoubtedly existed there before then but had not been noticed by butterfly experts.

The following year, TNC estimated that 1,392 Oregon silverspots were fluttering around the Cascade Head meadows. Two years later the population had declined steeply, to an estimated 790 adults, but then the next year it rebounded to almost as many as in 1989. Large swings in annual populations are not unusual in insects like butterflies, so things seemed pretty normal. But in 1993, the population plummeted; that year the estimate was 184 adults. The numbers stayed low, around 300 individuals, in 1994, 1995, and 1996, and then in 1997 dropped again, reaching a low of 57 butterflies in 1998. Working from the hypothesis that the population crashes of 1993 and 1998 were due to declining availability of early blue violets, the larval host plant, TNC began experiments to understand how prescribed burning would affect the violets. They compared management treatments of fall burning or mowing of the Cascade Head meadows, and found that early blue violets love fires. TNC's Debbie Pickering and her colleague Dan Salzer reported in 2014 that there was "a tenfold increase in violet seedlings in the burned treatments compared to a 29 percent decrease in the control and no change in the mowed plots." Although their research also showed that burning didn't drive back the introduced invasive grasses, they increased the scale of burning because of its positive effect on violets. But further studies convinced TNC that burning alone would not accomplish their goal of restoring habitat for the silverspot. So, they tried another tactic: planting violets. From 2010 to 2014, 28,000 violets were planted.

But even before that, by 1998, a sense of desperation had apparently taken hold, and a decision was made to supplement the natural population of silverspots. A captive-rearing program using female butterflies collected from Cascade Head was initiated in partnership with the US Fish and Wildlife Service (USFWS), the Oregon Zoo, and Lewis and Clark College (and later the Woodland Park Zoo in Seattle, Washington). One hundred or so captive-reared pupae were released at Cascade Head in 2000, 2003, and 2005.

Then another hypothesis emerged: that the problem was not only the availability of violets in the area, but that the 1993 and 1998 population crashes had reduced the genetic variability in the local silverspots and that an infusion of genes from another population might help. So, females from another Oregon silverspot population, similar to, but somewhat genetically different from the Cascade Head butterflies as determined by mitochondrial DNA analysis, were used in the captive-rearing program, and from 2007 through 2011 more than 500 captive-reared pupae were released each year. The peak effort was in 2009, when 1,209 captive-reared butterflies were released—almost as many as in the wild population in 1989. This population augmentation was expensive and was always intended only as a temporary stopgap measure, and in 2012, when butterfly numbers at Cascade Head appeared to be back in the pre-crash range because of the artificial supplementation, TNC stopped the releases to test whether they had restored a self-sustaining population. Unfortunately, that year butterfly numbers were immediately back to where they had been between 1993 and 1998, before any releases, and they have continued to drop since then. The 2018 population of silverspots on Cascade Head was estimated to number a mere five butterflies.

It's hard to draw firm conclusions from this fascinating case, which shows how scientists and stewards at Cascade Head tried to figure out what to do to help the Oregon silverspot, racing against the clock as butterfly numbers declined—except that it's clear in hindsight that the efforts so far have not succeeded.

~

The taxonomy of the silverspots—that is, the way butterfly experts categorize them into species and subspecies based on external characteristics, but which is also assumed or supposed to reflect their evolutionary history—is as complicated as the plot of Shakespeare's *A Midsummer Night's Dream*. The Oregon silverspot's subspecific name—the third name in the scientific system for naming genus, species, and subspecies—is *hippolyta*. In Shakespeare's play, Hippolyta was the wife of Theseus, king of Athens; but that play played on the Greek myth in which she was the queen of the Amazons, daughter of Ares, the god of war, who did finally marry Theseus of Athens after the Trojan War.

So here is the simple description—fasten your seatbelts—of the taxonomic situation of the silverspot group, quoted from the 2001 recovery plan for the Oregon silverspot:

> Ten species have a complex, polytypic population structure with over 100 geographic subspecies. Eight species and 36 subspecies of *Speyeria* are found in the Pacific Northwest. The Oregon silverspot butterfly is 1 of 15 subspecies of *S. zerene*. Subspecies of *S. zerene* are clustered into five major groups that are genetically distinct but not genetically isolated; some interbreeding may occur. These include the *bremnerii* group in the Pacific Northwest west of the Cascade Range and on the Northern California Coast. . . . The Oregon silverspot butterfly is one of five subspecies in the *bremnerii* group.

Whoa! Even for me, with academic training in ecology and evolution, this sounds pretty complicated. It turns out that it isn't easy to look at the external features of this group of butterflies and figure out how the biogeographic pattern we observe today developed and evolved. The genus *Speyeria* is notoriously variable in wing pattern and color, and the large number of subspecies has been defined on the basis of this visible variation, even though visible differences between subspecies from different regions are sometimes greater than the differences observed among different species in the same region. The silverspots are "a challenge to butterfly systematists," as a recent scientific paper understated, and that is important because many of the species and subspecies are threatened or endangered. "Many *Speyeria* species have been in decline for the past 200 years across North America as a result of habitat loss and degradation," say Ryan Hill and his coauthors in a 2018 paper. "Conservation efforts are needed to combat these effects. This requires a robust understanding of *Speyeria* ecology and population structure, including distinctiveness of populations."

The issue of "distinctiveness of populations" is important to conservation for a couple of reasons. The fundamental reason is that evolution is an ongoing process that requires genetic variation, the raw material on which natural selection acts, as species are forced to respond to changes in the ecosystems they inhabit and constitute. Individual species within a group like the silverspots are thought to contain unique genetic variation, packaged up into closed reproductive units (species); subspecies within a species are assumed to contain a unique sample of the species' genes, adapted to the local conditions in the range of that subspecies. Thus, species, subspecies, or even some populations are sometimes called "evolutionarily significant units," or ESUs. The Endangered Species Act allows for the listing and protection of a "distinct population segment," or DPS, of a species, rather than having to list the whole species if it is not threatened everywhere. *Speyeria zerene hippolyta*, as a subspecies, qualified for listing as a threatened subunit of the species. The criteria for qualifying as a distinct population segment involve the population's discreteness, significance, and conservation status. Congress directed the US Fish and Wildlife Service and the National Marine Fisheries Service to use their authority to list distinct population segments under the Endangered Species Act "sparingly," while still encouraging the conservation of genetic diversity by maintaining populations of threatened species over a representative portion of their historic range.

Because the taxonomic picture for *Speyeria zerene*, and for the genus *Speyeria* in general, is so muddled, recent research has turned to modern tools of genetic analysis, such as analysis of genetic variation in the DNA itself, to try to understand the distinctness, discreteness, and potential evolutionary significance of butterfly populations in this group. To get a better understanding of silverspot DPSs and ESUs, in other words.

Just to get the flavor of this research, let's peek into a fine study by a student-faculty team from Lewis and Clark College in Portland, published in 2013 in the *Journal of Insect Conservation,* titled "A Molecular Phylogenetic Analysis of *Speyeria* and Its Implications for the Management of the Threatened *Speyeria zerene hippolyta.*" Anne McHugh and her coauthors noted that the evolutionary relationship of the Oregon silverspot to other *Speyeria* species and subspecies is not well understood, and they tried to determine whether genetic analysis supported the current morphologically based taxonomy of the *Speyeria* species complex, and also whether it supported the Oregon silverspot's designation as an ESU under the Endangered Species Act. "The taxonomic confusion characterizing *Speyeria* raises the question of whether these taxa are as genetically distinct as their protected status implies, and suggests that a genetic analysis of *Speyeria* could provide important information to help guide conservation efforts," they wrote.

The research looked at genetic variation in two mitochondrial gene regions and four nuclear genes in *Speyeria zerene hippolyta,* seven other purported subspecies of *Speyeria zerene,* and eight other supposed *Speyeria* species, several of which had multiple subspecies. Their conclusion? Except for only one of the nine species, "Our data provided little support for most of the species currently recognized for western U.S. *Speyeria,* including *S. zerene,* and even less for the many subspecies designations. These genetic findings stand in contrast to the morphological differences recognized by experts." In other words, the study did *not* find genetic support for considering the Oregon silverspot an evolutionarily significant unit, and the authors almost seemed apologetic, saying that the study "does not provide a tidy resolution of this complexity," and that their "results for this group were particularly perplexing."

But the Lewis and Clark team were reluctant to upset the applecart that federal "threatened" listing of the Oregon silverspot had set in motion. Although their genetic data called into question its taxonomic and evolutionary status, they politely and conservatively wrote,

> However, the pattern created by these data is not strong enough to override other evidence supporting *S. z. hippolyta*'s status as a distinct ESU. This group displays specific morphological, develop-

> mental, and ecological traits that McCorkle and Hammond (1988) described as adaptations to the salt-spray meadows and windswept headlands that characterize its coastal habitat.

As McHugh and colleagues mentioned, the world's experts on this butterfly, David McCorkle and Paul Hammond, described (in the 1988 paper) four unusual characteristics of the Oregon silverspot that seem to be adaptations to the coastal prairie environment and are not found in the closely related subspecies, Bremner's silverspot, from the warmer Willamette Valley to the east. First, *S. z. hyppolita*'s small body size and dark coloration seem likely to enhance body heating from solar radiation. It flies even in cool, cloudy, or foggy weather. Its larval development is slow, leading to adult emergence in late summer and early fall when coastal weather is warmest and driest, and when many nectar plants are blooming. Finally, the Oregon silverspot exhibits a wide range of individual variation in larval development rate and adult emergence. Like the diversity of life histories in juvenile Chinook and coho salmon in the Salmon River estuary, this life-history diversity of Oregon silverspots could create a portfolio effect that reduces the risk posed by year-to-year variability and unpredictability of coastal weather.

How can we reconcile the recent molecular genetic findings with the way butterfly experts have described the species and subspecies based on external traits? And, more importantly, what does this have to say about conservation of the Oregon silverspot and its kin? The Lewis and Clark team wrote that understanding the "boundaries" of evolutionarily significant units "has important implications for management actions." That is why the Endangered Species Act recognizes ESUs, after all. McHugh and her colleagues warned quietly that the Oregon silverspot butterfly "provides one example of a group for which ambitious conservation activities are underway even though little is known about the population processes and historical biogeography that underlie its current distribution." Is there a risk to that?

A report from some scientists with the US Geological Survey and US Fish and Wildlife Service, published in 2016, used molecular genetic techniques to examine the genetic diversity in existing Oregon silverspot populations. Their main purpose was to assess the implications of the genetic data for future translocations of butterflies from one population to another, and for the establishment of new populations—both of which were already taking place under the Oregon silverspot recovery plan. The research concluded that if natural Oregon silverspot populations are genetically differentiated and discrete, then moving butterflies from one location to another to supplement an existing population could potentially "swamp out" the native genetic diversity of that population and perhaps decrease its evolutionary resilience. That would be similar to the effect that salmon hatcheries may have had, in some cases, in swamping out stream-adapted runs by introducing hatchery fish, especially genetically different fish from another stream and run. The report notes the challenge posed by finding a discrepancy between traditional subspecies designations and patterns of molecular genetic diversity. It points out that a lack of diagnostic genetic differences between subspecies—as the Lewis and Clark group found—"may call into question the taxonomic designations applied to different silverspot butterfly subspecies and populations, which likewise may have implications for their protected status."

~

Ambitious efforts to restore the Oregon silverspot are under way at Cannery Hill, in the Nestucca Bay National Wildlife Refuge, only eight and a half miles (as the butterfly flies) north of Cascade Head. The USFWS has been restoring a patch of coastal prairie on an old dairy pasture on a hill above the confluence of the Nestucca and Little Nestucca Rivers here for eight years. Encroaching shrubs like salal were mechanically removed at the edges of the former pasture; tractors scraped corridors through shrubs and trees to link patches of meadow; herbicide treatments pushed back invasive nonnatives; prescribed burns thinned the grass thatch and prepared the soil; about sixty thousand violets were planted; and hundreds of pounds of native seeds were scattered. Oregon silverspots were not found here when the butterfly was listed as threatened, but in July 2017, caterpillars from adults captured on Mount Hebo and reared in captivity at Seattle's Woodland Park Zoo were released, and adult butterflies were spotted about a month later.

Mount Hebo, a 3,157-foot Coast Range peak in the Siuslaw National Forest, is topped by meadows that support what is thought to be the last viable, self-sustaining population of the Oregon silverspot. It is the current source population for butterfly restoration efforts like those at Cannery Hill. The

Mount Hebo meadows are a legacy of the 1845 Nestucca Fire and another large fire in 1910 that burned over the mountain and killed the old forest that once covered it. Violets probably expanded into the open meadows from adjacent rocky areas. Settlers grazed their livestock in the meadows in the early 1900s, keeping them open.

Recent studies show that all of the Mount Hebo silverspots have the same mitochondrial genotype, genetic evidence that suggests a single ancestral female could have founded the population sometime after the Nestucca Fire, having been blown the ten or so miles from a coastal meadow on a strong west wind. A fertilized female can lay more than two hundred eggs, so she could potentially found a new population by herself. Debbie Pickering and other Nature Conservancy staff studied the dispersal ability of silverspots in 1998; they marked adult butterflies with fluorescent dust and found movement between sites five miles apart even with a very small sample size and limited period of observation. They concluded that dispersal, especially from north to south with the prevailing winds of summer, could occur with a fairly high frequency. Silverspot populations that were historically found along the coast were much more genetically diverse than the one now found on Mount Hebo, and it is ironic that those populations are now almost completely extinct, while the genetically impoverished population on Mount Hebo is being tapped to reestablish them.

The silverspots, genus *Speyeria*, are evolutionarily innovative and ecologically resourceful. One of those species, the zerene silverspot, *Speyeria zerene*, has spun off fifteen named subspecies, a complex group so similar in appearance that the genetic and taxonomic boundaries between them are blurry. They've been flapping their way over the west, tapping into the pool of genetic variation they carry to adapt to the grassland habitats where violets live. Loose on the landscape, they exist as metapopulations: networks of small, disjunct populations occupying a patchwork or mosaic of habitats between which they may move and interbreed. If the disturbances that maintain the prairie patches are frequent enough and the patches not too far apart, a population that becomes locally extinct can eventually be refounded from a nearby population.

Humans seem to be at least partly responsible for creating the habitats that the silverspots exploit. If Oregon coastal prairies can be called the First People's Prairies, then perhaps the Oregon silverspot, dependent on those salt-spray meadows, could be the First People's Butterfly. Burning coastal headlands is said by some to have been a ritual practice. Historian Stephen Dow Beckham says that "in accounts written at the mouth of the Rogue River in 1853–54 white settlers explained that the Indians there set fire to the headlands each year, believing that in doing so the salmon would return again to that stream." The Confederated Tribes of Grand Ronde sent a tribal fire crew to help TNC with the prescribed burn we were reseeding. Maybe this burn on Cascade Head could also be viewed as a ritual plea to the Oregon silverspot to please come back.

But Shakespeare's words from *A Midsummer Night's Dream* warn us not to get too attached to something ephemeral, and a subspecies of a rapidly evolving butterfly group is indeed ephemeral in an evolutionary sense. We poor humans shouldn't get too attached to it, lest we find ourselves playing god. What matters is the evolutionary creativity and resilience the Oregon silverspot represents.

~

Across two acres of slope, we spread a dozen cups of the long native grass seeds, trying for the one seed per square foot recommended by the experts. Such a paltry scatter, so few fragile seeds, each an almost-hopeless risk on the blackened bare soil.

For what?

Scattering seed to say we want to care for this place, make it whole again if we can, if that is even possible. Scattering seed that will wait for the rains and spring up green in our minds, putting roots into the future, holding onto the soil of hope. We dream of early blue violets thriving among the burnt blackberry canes and charred coyotebrush, and prickly caterpillars gobbling their leaves.

So long, silverspot, sail on. See you again, soon, I hope.

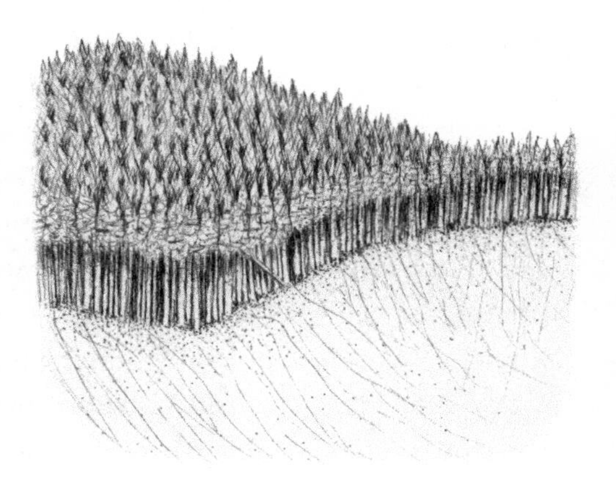

9 *Fish in the Forest*

Their plan wasn't to replace the forest, but to eliminate it and have plantations instead. It was to be closer to a cornfield than a forest.

—Andy Kerr, quoted in *The Final Forest* by William Dietrich, 2010

We spent the morning wending our way on wet forest roads up Drift Creek from its mouth on Siletz Bay to its source, where springs trickle from a steep hillside of volcanic rock. It was a bright, clear day, and from the head of the watershed, at an elevation of around 3,000 feet near the Coast Range divide, we could see Cascade Head and its grassy slopes above the Salmon River about twenty miles away. It looked tiny and vulnerable—familiar, and somehow precious. The foreground of the view was a quiltwork of recent clear-cuts and tree plantations in various stages of regrowth on privately owned timberlands. A little farther away, on US Forest Service land, the forest was thicker, older, greener. Our view took in the watersheds of the Salmon River, Schooner Creek, and Drift Creek, the most important watersheds for people and fish along this stretch of the Oregon Coast. The water trickling out of the rock behind us, and from similar springs and rills all along the Coast Range to the north and south, fed them all. Seeing this view, I felt a bit sick, and also a little angry. But deeper still was a longing for reconciliation—how do we heal this landscape?

I had met Conrad Gowell, my guide on this tour of the Drift Creek watershed, at the Otis Cafe early that morning. There was frost on the grass at the Sitka Center when I left there on the way to Otis; in the clear dawn light the roads looked slick, and I drove slow and easy. Conrad is the fellowship program director of the Native Fish Society, a nonprofit organization whose website says it is working toward "a future with abundant wild fish, free-flowing rivers, and thriving local communities," through science and advocacy "to cultivate the groundswell of public support necessary to revive abundant wild, native fish across the Pacific Northwest." He was about my son Jonathan's age, early thirties I guessed, with a thin dark mustache and short beard. While we waited for our food, Conrad spread out a detailed map and started to get me oriented. He probably knows the area better than anyone, having grown up fishing and hiking there with his grandfather, who loved the creek. It wasn't very long before the food came—for me a white cheddar omelet, hash browns, and the famous molasses toast—and we had to roll up the map.

After breakfast we started on our bottom-to-top ascent of the watershed. Where the creek flows into Siletz Bay it spreads out into meadows, formerly salt marsh but diked for dairy pasture in the 1920s and 1930s. The fog was just burning off and frost melting, sparkling the grass. We stopped where South Drift Creek Road crosses the river to see the cultural fishing site of the Confederated Tribes of Siletz Indians, the only tribal fishing site on the Oregon Coast where traditional dipnet and spearfishing is permitted by the Oregon Department of Fish and Wildlife (ODFW). It was a short stretch of the river near the bridge, a very difficult spot to fish. The tribe has caught only a few salmon there in the past few years, although the ODFW permit would allow them to take up to a hundred fish a year. Across the creek on the north side was a small pumphouse, from which Lincoln City can draw water from Drift Creek for its municipal water supply system. I'd already heard about the legal fight that took place from around 2003 to 2005, as Lincoln City acquired that water right to supplement its dependence on water from Schooner Creek.

Before the road had really started to climb up into the hills, we saw a white twin-cab pickup parked at a gate into some private land, with lettering on the door that identified it as from the Kernville–Gleneden Beach–Lincoln Beach Water District. The KGBLB water company supplies around two thousand customers. They received an award in 2001 for "Best Tasting Surface Water in Oregon," their website proudly proclaims. Conrad got out, introduced himself to Vern, the water company technician, and soon got permission for us to follow him through the gate to their water intake on lower Drift Creek. In his fluorescent orange sweatshirt, camouflage baseball cap, and baggy blue jeans, Vern seemed happy to have someone to chat with on this chilly morning. Vern's mission was back-flushing the intake pipes buried in the gravel along a stretch of the creek by pumping air into them to push out leaves and sediment that were starting to clog the intake holes. When he flipped a switch in the pumphouse, the big electric air pumps blasted into deafening action, and soon patches of bubbles were burbling up in the fast-flowing blue water of the creek.

We thanked Vern and said goodbye, and after another half hour of driving, Conrad turned up a road that crossed onto private commercial timberlands and the patches of clear-cuts started. To the east, and up another thousand feet, was a peak with a triangular clear-cut on the north side, white with new snow. I realized I'd seen that distinctive landmark from a kayak in the Salmon River estuary when I was looking for the tsunami layer in the salt marsh. My imaginary eagle's eye view of the landscape was finally closing a loop: I'd seen the watershed from bottom to top, and now, from top to bottom.

The road wound up through a mosaic of plantation forest at all stages, from the gray patches of last year's cuts to swaths of skinny, tight-packed trees ready to be harvested. Rotations here are thirty to forty years, depending on what the wood will be used for. On a steep slope across from the road, above a deep ravine, big old stumps stood out like scattered gray monuments in the graveyard of the original forest. Planted between them were four-foot-high Douglas-firs, like a Christmas-tree farm. We stopped to take in the view where the road ran up a ridge with a clear-cut on one side and a ready-to-cut crop on the other. On that side, the trees were closely packed, their six- to eight-inch trunks about six feet apart. Their branches closed off the sky, and it was very dark on the forest floor, which was dry and brown—no ferns, no moss, no green. I've crawled into a modern cornfield, and I can confirm that it is dark under that kind of corn canopy, too, not enough light for weeds to grow even if they hadn't been sprayed down. Of course, here in these tree plantations, the red alders had been sprayed down like weeds too. Alders are not tolerated in industrial timberlands, any more than monarch-butterfly-supporting milkweeds are tolerated in Iowa cornfields.

What happened? From the 1950s to the 1990s, federal, state, and private forest managers came in and "mined" old-growth forests for their incredibly high biomass of wood, and then converted them into "cornfield" forests. "Old-growth forest logging was questioned no more in mid-twentieth century Oregon than was slavery in mid-nineteenth century Mississippi," Andy Kerr wrote in *A Brief Unnatural History of Oregon's Forests*. Then came the "civil war" in forest management that eventually led to the "emancipation proclamation" of the Northwest Forest Plan in 1994. But taking in the view from the top of the Drift Creek watershed, it looked as if the "South" had won. This landscape spread out below us was divided, not united, with completely different moral standards regarding how to treat the land. In the foreground, on private forest lands, were monoculture tree plantations; in the distance, on National Forest lands, the clear-cutting had stopped and regrowth and restoration had begun.

Can a landscape thus divided against itself permanently endure, half slave and half free?

~

"Forest ecosystems interact in many and important ways with hydrological processes; indeed, it can be argued that the role of forests in watershed protection may be the most important category of ecological service provided by forests in the twenty-first century," say Jerry Franklin and his coauthors in their 2018 textbook, *Ecological Forest Management*. That's a pretty broad and bold statement about the interaction of water and forests, and we would not expect them to make that claim without solid evidence. The evidence comes from research at the H. J. Andrews Experimental Forest, located in the Willamette National Forest on the western slope of the Cascades east of Eugene, Oregon. The Andrews Experimental Forest is a sister to the Cascade Head Experimental Forest, if you will. In fact, it was once a sister biosphere reserve also; like Cascade Head, it was among the first crop of US biosphere reserves established in 1976. But in 2015, when UNESCO demanded a periodic review from all US biosphere reserves, the Andrews, unlike Cascade Head, opted to drop out of the Man and the Biosphere Programme rather than update their status.

In 2016, Timothy Perry and Julia Jones published a paper in the scientific journal *Ecohydrology*, based mainly on research at the Andrews, titled, "Summer Streamflow Deficits from Regenerating Douglas-fir Forest in the Pacific Northwest, USA." That might not sound provocative, but the first few sentences of the abstract pack a carefully worded scientific punch with important implications, and it's worth reading the original:

> Despite controversy about effects of plantation forestry on streamflow, streamflow response to forest plantations over multiple decades is not well understood. Analysis of sixty-year records of daily streamflow from eight paired-basin experiments in the Pacific Northwest of the United States (Oregon) revealed that the conversion of old-growth forest to Douglas-fir plantations had a major effect on summer streamflow. Average daily streamflow in summer (July through September) in basins with 34- to 43-year-old plantations of Douglas-fir was 50% lower than streamflow from reference basins with 150- to 500-year-old forests dominated by Douglas-fir, western hemlock, and other conifers.

First, these scientists admit that the ecohydrology of plantation forests needs more research. However, they then confirm that research clearly shows a dramatic impact of tree plantations on groundwater and streamflow—a reduction of up to 50 percent that lasts decades. After reading that study, one image from my trip up to the top of the Drift Creek watershed made more sense—the dry, brown, moss-and-fern-less forest floor under that canopy of dense young plantation trees. Young trees like that, growing fast, suck up so much water that not much can escape their thirsty roots, trickle into the soil, and eventually flow down through the ground to the streams.

The implications are potentially serious. Mature and old forests appear to be very efficient in water use, and generally produce more stable streamflows compared with forest plantations—although it gets a bit complicated when the age of plantation stands is taken into account, because immediately after logging in experimental watersheds, low flows can increase briefly, until the new crop of young trees gets big enough to really start sucking up water. Summer low-flow deficits following logging and replanting appear to last for at least forty to fifty years. The research suggests that because of the logging history of the entire Pacific Northwest region and the tiny amount of old forest left on the landscape, most watersheds in Oregon and Washington are probably experiencing severe, but previously unrecognized, streamflow deficits compared to pre-logging conditions. Since the hydrological effects are caused by the physiology of young trees, it isn't likely that changes in industrial forestry practices can reduce their harmful effect on streamflows. The only solution would seem to be reducing the area of forestry plantations and letting larger areas return to mature and old forest conditions.

Perry and Jones pointed out another obvious implication of their ecohydrological research: "Reduced summer streamflow has potentially significant effects on aquatic ecosystems. Summer streamflow deficits in headwater basins may be particularly detrimental to anadromous fish, including steelhead and salmon, by limiting habitat [and] exacerbating stream temperature warming." Because summer streamflow deficits may threaten fish like coastal coho salmon that are listed under the Endangered Species Act, plantation forestry practices may in turn exacerbate complex and contentious trade-offs between in-stream flows mandated to protect fish and human water needs, such as for drinking water and irrigation. If you are thinking that this science sounds politically sensitive in a state like Oregon, that's an understatement. A forest scientist familiar with this research suggested that I include the caveat that "scaling up to larger watersheds is as yet speculative, as is the connection to fish. Further study is needed." Even scientists sometimes tiptoe carefully around the forestry industry, which often, directly or indirectly, funds a lot of forest research.

Dr. Bob McKane is an ecologist with the US Environmental Protection Agency's (EPA) Western Ecology Division, based in Corvallis. He is leading an interdisciplinary group of scientists in the development and application of a landscape-scale ecohydrological model for watershed restoration planning called VELMA, an acronym for Visualizing Ecosystem Land Management Assessments. In developing the VELMA model, the EPA team started with data from small headwater catchments at the Andrews Experimental Forest, building on research by Perry and Jones and other Andrews scientists. VELMA estimates the effects of land-use and climate change scenarios on water quality and quantity, food and fiber production, fish and wildlife habitat, carbon sequestration, and other ecosystem services. It can help decision-makers and the public better understand the trade-offs among water, forests, fish, and a range of human benefits from different land-management scenarios.

VELMA shows how the conversion of old forests to young industrial forest plantations leads to higher evapotranspiration (water loss from soils through vegetation), lower low flows in streams, higher stream temperatures, higher peak flows (floods), more sediment, and less large woody debris that helps create spawning habitat for salmon. McKane and his colleagues have worked extensively with the Nisqually Tribe of Washington, modeling the watershed of the Mashel River, a tributary of the Nisqually River, which drains the western flank of Mount Rainer and flows into Puget Sound near Olympia. The Mashel watershed is approximately eighty square miles, very similar in size to the seventy-five-square-mile Salmon River watershed. Like the Salmon River, the headwaters of the Mashel have lots of timber plantations on forty-year rotations. Recently, the McKane team has been applying VELMA for salmon recovery planning, including in the Tillamook Bay watershed, not far north of Cascade Head and the Salmon River.

In order to scale up the effects of forest practices in the VELMA model, its developers used Landsat satellite data in a program called LandTrendr to create time-lapse animation of landscape-scale ecological change. They call it "geospatial time travel." Google Earth Engine is making a similar kind of time-series satellite imagery more and more available also. Its time-lapse feature "is a global, zoomable video that lets you see how the Earth has changed over the past 35 years," according to their website. A few days after exploring Drift Creek with Conrad Gowell, I decided to do a little geospatial time traveling myself and have a time-lapse look at the Salmon River–Schooner Creek–Drift Creek ecosystem from space using this Google tool. It was a mind-boggling ride—a new kind of "eagle's view," from the back of a Landsat Google-eagle.

~

Salmon have what is known as an anadromous life cycle: they hatch from eggs in freshwater streams, move into the ocean after feeding and growing as juveniles, grow and mature after one to three years at sea, and return to their natal streams to spawn as adults. Their fairly, but not completely, faithful return to the stream or river where they hatched, guided by "olfactory memory"—essentially being able to smell, or taste, the unique terroir of that home stream and follow it—has resulted in strong local adaptation and genetic differentiation within a given species of salmon.

Unfortunately for salmon, returning to streams and rivers to spawn makes them extremely vulnerable, at that stage of their life cycle, to being caught and eaten. Native Americans arrived in the Pacific Northwest at least ten thousand years ago, and archaeological evidence indicates that intensive salmon fisheries began on the Columbia River eight thousand to nine thousand years ago. From the time sea levels stabilized after the last Ice Age, around five thousand years ago, Native American coastal cultures developed greater and greater dependence on salmon. Despite this, however, cultural mechanisms that maintained salmon populations by restricting fish harvests to sustainable levels evolved in these tribes.

In contrast, European cultures never seemed to manage salmon sustainably. In the Atlantic, generally unsuccessful attempts to conserve salmon date back many centuries. The Magna Carta, signed in 1215, required that the king's fish weirs be dismantled to protect salmon runs for public use. A law against taking salmon out of season or from private waters without permission was passed in England in 1285, and another forbidding the construction of barriers to fish passage was approved in Scotland in 1318. But by the 1850s, economic forces had overwhelmed the good intentions of the English and Scottish laws and regulations regarding harvest levels and fish passage, and clearing of forests and channelization of rivers destroyed salmon habitat. Soon wild salmon in the eastern Atlantic were hanging on only in a few Scottish rivers.

In New England, the story was repeated. There, as in the Pacific Northwest, native cultures had depended on salmon for thousands of years, but with the arrival of European colonists, overfishing, forest clearing, and mill dams that blocked fish passage rapidly depleted salmon runs. Salmon once ranged from Long Island Sound to Hudson's Bay, and spawned in all of the major rivers, but by the mid-1700s they were scarce south of Maine. And the same story was repeated in the Pacific Northwest, starting about the time of the California Gold Rush in 1849.

Salmon and humans, at least the modern European version of our species, have had a hard time coexisting over the last few centuries. Overharvesting was the initial whammy, depleting runs in England, Scotland, New England, and the Pacific Northwest. Nets strung across rivers, and other intensive fishing technologies, often led to harvest of 90 percent of the annual population of spawners. Unsustainable harvesting—fish "mining," really—was often driven by "commodification" and long-distance trade. Canning technology played a big role in turning salmon into a commodity and depleting local runs. Commercial salmon canning started in Scotland in the 1830s, spread to Maine a decade later, and to California in the 1860s. After salmon runs in northern California dwindled, the fish-mining canneries moved to the Columbia River in the 1880s, and then on to Puget Sound. Global export of canned salmon within an essentially unregulated fishery caused the first dramatic collapse of Pacific salmon populations. But overharvesting was soon compounded and overtaken by habitat loss. Log drives in rivers scoured out many spawning streams. In the Columbia River Basin, construction of the big hydropower dams began during the Depression and accelerated around the time of World War II.

The prescription for plummeting salmon populations was always, "Don't worry, we'll set up hatcheries to dump more baby fish into the rivers and ocean! We'll fool Mother Nature! Don't worry." It just didn't work, especially after overfishing had hammered the genetic and life-cycle diversity of the fish, and their habitat was on the ropes. In the Columbia Basin, which is thought to have supported annual runs of up to sixteen million salmon prior to European settlement, runs are now down to between one and two million fish, 80 percent of which are released from hatcheries. In a little more than 150 years, populations of many West Coast salmon have declined to the point where they are in danger of extinction from the deadly combination of overfishing, loss of freshwater and estuarine habitat, hydropower development, hatchery practices, and variable ocean conditions.

Twenty-eight stocks, or "evolutionarily significant units," of Pacific salmon have been listed for protection under the federal Endangered Species Act by the National Marine Fisheries Service. Evolutionarily significant units, or ESUs (as explained in the essay on the Oregon silverspot butterfly) are thought to represent an important sample of the genetic diversity of the species and to be substantially reproductively isolated from other populations.

They are considered the equivalent of "species" under the Endangered Species Act. One of those twenty-eight listed ESUs is the Oregon Coast coho salmon.

Oregon Coast coho salmon (*Oncorhynchus kisutch*) inhabit coastal rivers south of the Columbia River and north of Cape Blanco. Between one and two million fish are thought to have spawned in these rivers annually before European settlement, but they began to decline in the mid-1900s, and by 1983 the population had dropped to fewer than fifteen thousand fish. The runs then improved for a few years, but dropped again in 1990. Oregon Coast coho were finally listed as threatened in 1998, and remain so today. The Oregon Department of Fish and Wildlife estimated that 57,000 coastal coho spawned in 2015, some improvement from the 1980s and 1990s. Fairly dramatic fluctuations in the population from year to year are speculated to reflect differences in ocean conditions and marine survival, according to the National Marine Fisheries Service's recovery plan for the Oregon Coast coho.

When a species is listed under the Endangered Species Act, "critical habitat" needed for its protection and recovery must be identified and designated. Critical habitats are the areas within the geographic range of the ESU that have physical or biological features essential to the survival and recovery of the group, and special management may be required to protect those areas. In its "Oregon Coast Coho Conservation Plan," ODFW has developed detailed maps of habitat quality and accessibility for coho salmon, including for the Salmon River, Schooner Creek, and Drift Creek.

Protection and restoration of critical habitat in watersheds used by threatened salmon stocks began in Oregon in the late 1980s, when the state legislature established the Governor's Watershed Enhancement Board, later renamed the Oregon Watershed Enhancement Board—OWEB as a lot of people call it, using its acronym. In 1997, one year before Oregon Coast coho were listed, the ODFW developed the Oregon Plan for Salmon and Watersheds and began funding local watershed councils to help implement it. The councils were supposed to provide a local level of stakeholder input and collaboration, below the federal and state levels.

One of the watershed councils established in the late 1990s was the Mid-Coast Watersheds Council, which covered a significant part of the range of Oregon Coast coho salmon. A sub-basin planning team focusing on the

Salmon River, Schooner Creek, and Drift Creek watersheds decided that the issues there were sufficiently unique, and that there was enough volunteer energy and agency support, to justify an independent Salmon Drift Creek Watershed Council, which was established in 2005. Since then, it has received grants and donations from a variety of public and private sources and worked with city, county, state, and federal agencies, and private landowners, on many monitoring and restoration projects in the watersheds in its jurisdiction. It worked closely with the Forest Service to produce the Cascade Head Biosphere Reserve Periodic Review in 2016.

David Montgomery, in his book *King of Fish,* wrote,

> Salmon are resilient, robust animals that can rapidly colonize new environments. They are more like weeds than like a sensitive bird that can only nest in a special type of tree that occurs in a particular type of forest in a couple places on earth. Even so, we are managing to drive them to the verge of extinction across much of their range. . . . Given half a chance they can take care of their own existence quite well and expand to fill the available habitat. But for a century and a half we have sustained a pace of landscape-scale changes that salmon have never experienced before except over short time periods and across limited portions of their range. By disturbing everything everywhere all at once, we risk leaving them no sanctuary from which to repopulate depleted rivers and streams.

Now, with effort and funding from federal agencies like the National Marine Fisheries Service and Forest Service, state agencies like ODFW and OWEB, and local organizations like the Salmon Drift Creek Watershed Council and the Native Fish Society, we are trying to repair the damage we have caused and restore watersheds and salmon. My bet is on the salmon, if we give them half a chance.

~

Almost seventy-five people filled the dozen round tables set up in the Yachats Community Center—an old school gym converted to a meeting space—finishing up their plates of enchilada casserole and salad before the meeting of the Mid-Coast Water Planning Partnership began. The walls were hung with flip-chart sheets with notes from past meetings and graphs of water flows in the rivers of the middle Oregon Coast. People milled and chatted before the facilitators called the meeting to order. These people knew each other, mostly, from more than a year of work together in this forum, organized by the Oregon Department of Water Resources, the state water management agency. The participants were water system managers from all the towns along the central Oregon Coast, from Lincoln City to Yachats. They were rural farmers who depend on water rights for irrigation and rural residents who use water from wells for their domestic consumption. They were ecologists and conservationists and tribal representatives concerned about leaving enough water in streams for salmon and the healthy aquatic ecosystems fish need to survive.

The eight watersheds represented in Mid-Coast Water Planning Partnership are, starting from the north, Salmon River; Siletz Bay ocean tributaries; Siletz River; Depoe Bay ocean tributaries; Yaquina River; Beaver Creek ocean tributaries; Alsea River; and Yachats River. The "ocean tributaries" refer to areas of the coast with outflows from short, small watersheds that never amalgamate into a "river," with several together often forming bays at the coast. Many of these little rivers are called "creeks"—like Drift Creek, which is one of the Siletz Bay tributaries.

The hydrographs posted on the walls underlined the purpose of this planning group. I had never really thought before that western Oregon, reputed for rain and in a "temperate rain forest" ecosystem, could get thirsty. But it can, and does, at the end of the summer dry season. Even though the Oregon Coast is very wet on average, that wetness is strongly seasonal. In fact, during the summer, it is very dry—almost no rain falls here from May through September, while up to a hundred inches or more, on average, fall during the other seven months. Around 90 percent of the precipitation falls from October to April, and 10 percent during the other five months. The short and small rivers that flow to the mid-Oregon coast from the divide of the Coast Range tap only that seasonally variable precipitation, so of course the graphs on the walls showed a strong peak in the wet season and a valley in the dry months. Low-season flows, in August and September, were somewhere around ten times lower than the winter flows of February, March and

April. But even though the flows are strongly correlated with precipitation, there was that base flow in the creeks and rivers during the late summer, after months of almost no rain. Somehow the land was retaining some water and releasing it with a time delay—that's where the ecohydrological function of forests explains some things. And, based on the research, that's where forest restoration could put the streamflow-enhancing effect of older forests back on the landscape.

Some years are drier than others, and climate change may be making them drier. The year 2015 was a very dry year in Oregon. Streamflows were exceptionally low, and water temperatures therefore high, reaching levels that threatened cold-water-loving salmon. Flows in many coastal rivers were as low or even lower in 2018. The City of Yachats put water conservation plans into effect in both of those years. Peak water demand up and down the coast coincides with low late-summer flows, and water demand is projected to outstrip available supply within fifty years in many coastal communities, especially if the climate gets warmer and drier.

A working group focused on "instream" flows and water rights made a presentation about their work and progress. In Oregon, instream water rights are held by state agencies for a public purpose, such as recreation, pollution abatement, navigation, and the maintenance and enhancement of fish and wildlife populations and their habitats. Only three state agencies can hold these rights: the Department of Fish and Wildlife, Department of Environmental Quality, and Parks and Recreation. In a way, instream flow rights are a water right for the river itself, and for the public benefits it provides. They are a management tool for balancing the water needs of humans and other water-dependent species. Of all of the rivers in the mid-coast region, two—Salmon River and Schooner Creek—have a gap between the allotted instream water rights and the average flow in August and September, when there is less water in the river than fish need.

At the end of the meeting in Yachats, I wondered, What are we going to do about it all?

~

There are many paths toward a more harmonious and healthy balance of people, forests, and fish in this coastal landscape. Bob McKane and his colleagues used the VELMA model to help envision a way forward. They modeled four main types of benefits from forest landscapes—forest products, local income from those forest products, salmon habitat quality, and ecosystem carbon stocks—under three alternative management scenarios. In the status quo scenario (industrial forestry plantations), forest product production and profits to investors in the industry were maximized. But because most of those forest products are exported without much processing or value added—as raw logs or chips, for example—local forest-based jobs and income are not very high. Salmon habitat quality is low, continuing to threaten the recovery of wild populations, and forest carbon stocks are only moderate. In a second scenario, the no-cut, bring-back-the-old-forest scenario, forest products and local jobs are basically zero, but salmon habitat quality and carbon storage—and certainly forest biodiversity—is maximized. In a third scenario, labeled "multi-stakeholder community forestry," ecological, economic, and cultural benefits for local stakeholders are balanced and optimized. In that scenario, forest products production is reduced to about one-third that of the current cornfield-forests scenario but is still significant, and local forestry jobs and income are around three times higher because much more value is added at the local level through processing and manufacturing. Salmon habitat quality approaches that in old forests in this scenario, and carbon storage is high.

Looking at these VELMA simulations, the way forward seems like a no-brainer. The multiple-value, multiple-stakeholder management vision seems like it could provide community jobs and income, drinking water, salmon recovery, flood protection, biodiversity conservation, recreation, and carbon sequestration.

This multiuse forest management is a middle ground between status quo commercial forestry and a pure old-growth restoration vision. In their book *Ecological Forest Management*, Jerry Franklin and his colleagues describe what they call "shades-of-green" landscapes, in which "all patches are consciously managed to sustain elements of structure, function and biodiversity." Such landscapes would look and function very differently than the current landscapes of the Pacific Northwest, which are still strongly partitioned into private industrial forestry lands for intensive wood production and public lands more and more devoted to conserving biological diversity and providing ecosystem services. A multiuse shades-of-green landscape model fits perfectly into the biosphere reserve vision of a sustainable, resilient balance between people and nature in every place and region. It would finally end the civil war that has left the ecological landscape half slave and half free.

Maybe it's just my congenital optimism, but I can imagine standing where I did at the headwaters of Drift Creek and looking out over a shades-of-green landscape, Cascade Head still far off in the distance, but this time not looking so vulnerable. The large wounds of clear-cuts will mostly be gone, the plantations growing older and being ecologically managed for diversity of age, structure, and function. The timber management investment organizations (TIMOs) and timberland real estate investment trusts (REITs) of the current "Wall Street forestry" era will be a thing of the past, and instead we'll have ESIMOs, ecosystem services investment management organizations, and EITs, ecosystem investment trusts. They will invest in watershed restoration, converting "cornfield" forests back to older, biodiverse, climate-change-resilient forests that can provide water, fish, wood products, jobs, recreation, and climate-buffering carbon sequestration. My bet is that, in a climate-changing, economically evolving world, all of those things—but especially water—are going to be more important than cheap two-by-fours and toilet paper.

Your bet?

10 *Beavers in Pixieland*

Beaver have great fun while growing up.

—Enos Mills, *In Beaver World,* 1913

Pixieland was the short-lived dream of Jerry Parks, owner of the popular Pixie Kitchen restaurant in Lincoln City, who, in 1965, purchased a fifty-seven-acre triangle of salt marsh between the Salmon River, US Highway 101 and Oregon Highway 18 to build an amusement park. To develop the site, the entire area was surrounded by a dike, a tide gate was installed on Fraser Creek, and the marsh was filled to create a dryland building surface. Pixieland Amusement Park opened in 1969, dedicated by Oregon's then governor Tom McCall. There was a roller-coaster ride called The Log Flume, where visitors rode in cars shaped like hollow logs and splashed through stretches of water. A tiny narrow-gauge train called Little Toot ran around the perimeter on the dike. Melodramas were offered at the Blue Bell Opera House. Visitors could enjoy scones from Fisher's Scone House, whose plaid roof was built to resemble a traditional Scottish tam o'shanter cap; indulge in malts, shakes, or ice cream at the red-roofed Darigold Dairy Barn; or have a sandwich at the Franz Bread Rest Hut, a building shaped like a giant hollow log with a huge loaf of bread perched on top. To support all the fun, there were roads, a parking lot, an RV park, and a sewage treatment plant.

Pixieland was exactly the kind of development that people pushing for the protection of the natural and scenic values of the area hated. Its development served as a catalyst for the political action that eventually created the Cascade Head Scenic Research Area. Jerry Parks's memorable defense of Pixieland at the Forest Service's public hearing in Lincoln City in 1972—when he said, "We can't pay our taxes with salal bushes!" and warned about the creeping influence of "Big Brother" (the US government)—couldn't overcome the tide of sentiment for controlling tourism and development in the area. By 1974—the same year the Cascade Head Scenic Research Area Act was passed, but not because of it—Pixieland was bankrupt and closed.

The Forest Service purchased the Pixieland property under its CHSRA mandate in 1981. The buildings were torn down, leaving behind the concrete foundations, paved roads, parking lots, ditches, dikes, and tide gate. Invasive alien species—Himalayan blackberries, Scotch broom, and reed canary grass—took over the site. Finally, with fresh funding, further restoration work began in 2007 and was completed in 2011. First the concrete building foundations and asphalt roads and parking lots were torn out. Then 27,000 cubic yards of fill was removed to restore the marsh surface to its original level. Finally, almost half a mile of dikes were removed and another half a mile of ditches filled to restore natural tidal flows to the area.

That morning I was standing on the shoulder of Highway 101, looking east across the healing landscape where Pixieland once sprawled and where Fraser Creek was reconnected to its old channel in 2017. I'd been told by Forest Service hydrologists that a little farther up the creek there was recent evidence of beavers. The morning was half-sunny, not too windy, and fairly warm, so I decided to go explore a bit and see if I could find some beaver-works. I drove to one of the old entrances to Pixieland along Oregon Highway 18, now closed by a Forest Service gate, and walked west on an old gravel road starting to be overgrown with blackberries. Tiptoeing through blackberry tangles and following elk trails, I tried to get close to Fraser Creek, but at each try I came to a dead end where water flooded the ground under alders and spruces. Finally, without wading, I found some fresh beaver cuts and a couple of small dams. They looked active and recently repaired, but they were nothing like the impressive stream-spanning structures I've seen in the Rockies and that I'd been imagining here. But here they were, subtly spreading water out from the creek into forest and marsh, the beginnings of a natural, long-term ecological restoration project. Beavers were back, playing in Pixieland again.

~

The Castoridae, or beaver family, evolved about forty million years ago during the Eocene Epoch. During the ice ages of the Pleistocene, giant beavers as big as bears wandered North American wetlands. The fossil record hasn't yet confirmed that they built dams and lodges like their modern relatives, but their huge incisors suggest that they would have been formidable lumberjacks. The giant beavers disappeared at the end of the last Ice Age, about eleven thousand years ago, around the same time as many other large mammals of the Pleistocene megafauna, such as mammoths, mastodons, horses, and gomphotheres. Early human hunters have been implicated in that extinction event, but it's not clear whether they had a hand in the demise of the giant beaver. The Castoridae once included many more species, but now only two remain: *Castor canadensis*, the North American beaver, and *Castor fiber*, the Eurasian beaver. American and Eurasian beavers are almost identical in appearance, but genetic differences support their distinctiveness as separate species.

Beavers are the second-largest and one of the most ecologically unique rodent species, but surprisingly little is known about their evolution. Recent genetic research using mitochondrial DNA—the same type of genetic analysis I've described for silverspot butterflies—suggests that North American and Eurasian beavers started to become separate species around eight million years ago, when their ancestors crossed the Bering land bridge between Siberia and Alaska. Since the estimates for the earliest appearance of *Castor* on both continents overlap, it's not entirely clear whether the ancestral beavers were moving from North America to Eurasia or vice versa.

Both species are large, semiaquatic rodents, with webbed hind feet, a flat scaly tail, and chisel-like incisors evolved for cutting trees and branches and gnawing bark. Both build dams of felled trees, branches, mud, and vegetation, and often construct lodges of similar materials. They live in family groups or "colonies," usually made up of around four to eight individuals. A typical beaver family consists of a pair of adults and their offspring of both the current and previous years. The two breeding adults normally raise three or four offspring, called "kits," each summer. Both male and female offspring disperse from their natal family at the age of two to find mates and found their own colonies. Groups of beavers may construct several dams. In favorable habitat they average one beaver colony per mile and build between one and five dams per mile of stream.

Beavers defend their stream territories and mark them with a secretion, called castoreum, produced from specialized gland-like sacs near their tails. Both male and female adults secrete and mark with castoreum, which is a complex concoction of aromatic organic compounds. Unfortunately for beavers, if they smell castoreum, they come to investigate who has been around, and trappers discovered that castoreum-scented traps easily catch them.

Beavers spread water out over the landscape with their dams, in the process often increasing their own food supply. Leaves and bark of water-loving trees like willow, alder, cottonwood, and aspen, and other kinds of aquatic vegetation, are beaver staples. Ponds and lodges also provide protection from predators, among which wolves and bears were historically most significant. Beaver dams create a cascade of ecological effects. By flooding riparian zones, they can kill trees and create wetlands; their ponds trap sediments and organic material, modifying decomposition and nutrient cycles; they increase the diversity of aquatic habitats, creating niches for more species. And their dams can wash out in floods, sending wood and sediments downstream, affecting habitat for fish and other aquatic species over long distances. Beavers have been called "ecological engineers" or "landscape engineers" because of the effects of their dam-building.

"The peculiar charm and fascination that trees exert over many people I had always felt from childhood, but it was John Muir, who first showed me how and where to learn their language," wrote Enos Mills in his 1909 book *Wild Life on the Rockies,* which he dedicated to Muir. He then confesses, "I have never been able to decide which I love best, birds or trees, but as these are really comrades it does not matter, for they can take first place together. But when it comes to second place in my affection for wild things, this, I am sure, is filled by the beaver."

Enos A. Mills was born on a Kansas farm in 1870, thirty-two years after Muir. His life story is so improbable and interesting, and so nearly unknown by most Americans, that it deserves a bit of attention, especially when talking about beavers. Mills left Kansas for the Colorado Front Range at the age of fourteen, seeking a cure for a mystery illness that some speculate was an allergy to wheat. At fifteen, he established his own homestead near Estes Park, climbed 14,255-foot Longs Peak for the first time, and fell in love with beavers. After a few years in Colorado, apparently not eating as much wheat as he had in Kansas, he felt healthy and moved to Butte, Montana, where he lived and worked intermittently between travels along the US Pacific Coast and to Alaska and Europe. In 1889 he happened by chance to meet John Muir on a beach in San Francisco and was so inspired by their conversation that he dedicated himself to nature writing and conservation. After moving back to Colorado in 1902, he started a nature tourism business, at his old homestead near Longs Peak, and thought of himself as Muir's disciple and acolyte, advocating for the conservation of nature and scenery in the Rocky Mountains as Muir did in the Sierra Nevada. As a nature guide, Mills climbed Longs Peak more than three hundred times, and took schoolchildren and groups of ladies in long dresses on hikes to watch beavers. He was a decent photographer and enthusiastic nature writer. In lectures, articles, and books, he led an effort to create Rocky Mountain National Park, lobbying nationally with help from the Sierra Club, which Muir helped found in 1892. In 1915, Rocky Mountain National Park became the tenth national park in the United States, a year after Muir's death. In a biography of Mills, the National Park Service calls him the "Father of Rocky Mountain National Park."

Rocky Mountain became a UNESCO biosphere reserve in 1976, at the same time as Cascade Head. It completed a periodic review in 2016, as did Cascade Head, and remains an active biosphere reserve. Enos Mills would, I'm sure, be proud.

Mills's fascination with beavers led him to spend a lot of time observing them, certainly more than any previous naturalist. One year he visited and observed the same beaver colony near Estes Park every day for more than two months, and his general observations span twenty-seven years. A chapter in his 1909 book *Wild Life on the Rockies* is titled "The

Beaver and His Works," and the 1913 book *In Beaver World* summarizes his lifetime of observations. One thing about beavers that delighted Mills was their play:

> Beaver have great fun while growing up. Posted on the edge of the house, they nose and push each other about, ofttimes tumbling one another into the water. In the water they send a thousand merry ripples to the shore, as they race, wrestle, and dive in the pond. They play on the house, in the pond, and in the sunshine and shadows of the trees along the shore.

Finally, at the end of a playful summer, the kits start to get serious lessons in tree-cutting and dam-building, he wrote.

Mills also had a solid understanding of the ecological role played by beavers in maintaining ecosystem services that were valuable to humans. "His engineering works are of great value to man," Mills said of the beaver. "If he and the forest had their way with the water-supply, floods would be prevented, streams would never run dry, and a comparatively even flow of water would be maintained in the rivers every day of the year."

~

It was the Eurasian beaver that first became a target of human commerce. Beaver fur turns out to have a fine underfur, or wool, with properties that make it the most desirable of any fur for making felt. And felt has unmatched properties for making hats, especially the wide-brimmed and shaped ones that became fashionable in the 1600s. Beginning then, European hatmakers placed a high demand on beaver fur, and *Castor fiber* was trapped nearly to extinction. But contact between European cod fisherman and Native American tribes on the coast of Canada in the 1600s identified a new source of beaver fur: *Castor canadensis*. Trade in beaver-pelt robes from North America initiated the French fur trade, with voyageurs ranging far into the Great Lakes area to acquire beaver skins. Much as with salmon, when beavers were hunted to commercial extinction in Europe, the trade moved to eastern North America. When they were hunted to commercial extinction there, the trade looked west.

Reports from the Lewis and Clark Expedition of 1804–1806 stimulated the westward push for new sources of beaver and other furs, and "mountain men"—fur-trapping-and-trading explorers like Jim Bridger, William Sublette, and Kit Carson—led the way across the Rockies to trap and trade with Native American tribes. The fur trade soon reached Oregon. Fort Astoria, called the "first US community on the Pacific Coast," was established in 1811 by John Jacob Astor's American Fur Company to compete with British fur interests, stimulating a geopolitical rivalry between England and the United States along the Columbia River. An overland expedition mounted in 1810–1812 to supply Fort Astoria discovered the South Pass route west of Fort Laramie that would later become the main pass over the Continental Divide on the Oregon Trail. Astor's company did not make money and was soon sold to the North West Company, another fur company that had been established in Montreal to try to break the Hudson's Bay Company's stranglehold on the fur trade. But the North West Company merged with the Hudson's Bay Company in 1821, and in 1825 they established Fort Vancouver at a site just across the Columbia from present-day Portland, with the aim of exploiting Oregon's beavers. As competition between the United States and Great Britain for control of the Pacific Northwest intensified, Hudson's Bay Company trappers deliberately created "beaver deserts," removing all beavers in the watersheds they trapped so that American fur trappers and the settlers who followed them would have less incentive to come to those areas.

Hats made exclusively from beaver wool were the most expensive and of the highest quality. Being able to afford a "castor," as a pure beaver-felt hat was called, made a visual statement about the wearer's wealth and social status. Just think about Rich Uncle Pennybags, the Monopoly Man: tall top hats, the most expensive of which were made from beaver fur, were a symbol of the upper class, capitalism, and the world of business. But fashion can be fickle. By the mid-1830s, at around the same time that beaver populations in Oregon were being destroyed by trapping, new silk hats became all the rage, and demand for high-topped beaver hats plummeted. In Oregon, the beaver trade was a "flash in the pan," as beavers were hunted to commercial extinction in only a few decades and beaver-felt hats went out of fashion.

The fur trade had involved Native Americans since its beginnings on the Atlantic Coast, and so it did in the Pacific Northwest. Native tribes often had more ecological knowledge of an area, and better hunting and trapping skills, than Euro-American trappers. And, of course, they lived there, and controlled the territory politically and physically. Trading furs for European trade goods like steel knives and axes, cloth, beads, cooking pots, and guns and gunpowder created a web of economic interdependency that required negotiations and had implications for political alliances between Native Americans and Euro-Americans. The collapse of the beaver trade in some ways weakened the power of Native tribes vis-à-vis Euro-American colonists and settlers.

Thus, "globalism," fashion, and capitalism were forces that began to impinge on the Cascade Head and Salmon River ecosystem and its Native American inhabitants two centuries ago—because of beavers.

It is estimated that there were sixty million to four hundred million beavers in North America prior to European settlement, according to an authoritative summary of beaver ecology by Robert Naiman and his colleagues. If anything, the wide range of that estimate suggests how little is still known about the ecology of the species. By 1900 beavers were "almost extinct," resulting in an estimated loss of around 88,000 square miles of wetlands, much of which would have been occupied by or created by beavers—an area about the size of the US state of Minnesota. In 1988, Naiman estimated there were six to twelve million beavers in North America; in 2018, Ben Goldfarb, in his book *Eager: The Surprising, Secret Life of Beavers and Why They Matter,* says that there are now around fifteen million, "though no one knows the number for certain."

~

Looking for beavers in Fraser Creek that morning reminded me of another beavering adventure, on the same trip to Ukraine when I visited the Askania-Nova Biosphere Reserve, described earlier, but in another part of that large and ecologically diverse country. Sergii Mykolayovych Zhyla, director of the Polisky Nature Preserve, a *zapovednik,* led the way through the sphagnum bog toward a stream to show us one of his proudest achievements—restoring beavers to the area. We were in northern Ukraine, not many kilometers from the border with Belarus, in an area called Polissya. It was an overcast afternoon in early April as we walked over the spongy ground through a forest of scraggly, dwarfed Scots pines (*Pinus sylvestris*), their orange bark adding a warm accent to the cool gray and green palette of the landscape. As we approached the stream, the pines gave over to white birches and willows and the ground got soggier; it was impossible to avoid sinking in water to the ankles even with carefully chosen steps. We heard the trickle of water before we saw the dam, which was an impressive structure of birch and willow almost as tall as a person. Its pond above it was a mirror of black water, dark from tannins leached from the bog vegetation, reflecting white-trunked birches just putting out a hint of spring-green leaves against the gray sky. Soft silver-furred catkins were popping out on the pussy willows, and on hummocks of high ground the heather was in pale-pink bloom.

Sergii Mykolayovych Zhyla looked like a cross between an academic biologist and a backwoods hunter. His sharp features and close-cropped salt-and-pepper beard were set off by a dapper flat cap of black felt; he wore a zippered fleece turtleneck under a camouflage jacket, blue jeans, and rubber boots. Zhyla was enthusiastic about his beavers. He had seen evidence that they were around, or passing through, he said, and to entice them to stay, he had constructed the beginnings of two small dams on the stream himself. That apparently did the trick, and like American beavers, who are rumored to be attracted by the sound of running water and determined to silence it, *Castor fiber* stayed and expanded Zhyla's incipient dams. Water is spreading out and rewetting the peat sponge.

Polissya is a flat landscape of boreal forests, lakes, meandering rivers, and wetlands—bogs, swamps, and marshes—a legacy of the Ice Age. The edge of the Eurasian continental ice sheet stopped right here, melting and leaving abandoned chunks of ice as it retreated northward under a warming climate. The ancestral Pripyat River flowed eastward along the ice front. Many northern species now reach their southern limit here. The area lies at something of a "continental divide" of eastern Europe: the Bug River flows from this area north across Poland and empties into the Baltic Sea, and the Pripyat River flows east along the border of Ukraine and Belarus, eventually turning south and merging with some other rivers to form the Dnieper, the main river of Ukraine, which flows south across the middle of the country into the Black Sea.

Geologically and ecologically, the Polissya landscape is very similar to the lake country of Minnesota, Wisconsin, and Michigan. It looks much like the area where John Muir's family homesteaded at Fountain Lake, Wisconsin, and like Aldo Leopold's Sand County along the Wisconsin River, which at one time flowed along the edge of the North American ice sheet. Both Muir and Leopold would feel at home in Polissya, a land of cranes and water.

During the Soviet era, a canal was built that took water out of a tributary of the Pripyat River near the Polisky Nature Reserve, which started to de-water the landscape around the preserve. But the soils in these peatlands are sandy and nutrient-poor, and the remote farms were abandoned without complaint when the *zapovednik* was created in 1958.

One of Zhyla's goals, besides protecting habitat for wildlife, is to retain and store water. Besides beavers, who can aid and abet the process of rewetting the landscape, his biggest ally is peat. "Peat moss" or "peat" is a common term for mosses of the genus *Sphagnum*. There are hundreds of species of *Sphagnum*, and most have evolved an amazing capacity for storing water in their cells—up to twenty times their dry weight or more—to enable them to survive dry periods. Forest fires are a threat to the ecosystem here. If the peat dries out, it can burn deep into the ground, making peat fires hard to extinguish. Fires really became a problem starting in 2002, when the climate started to become drier. "In all of history the swamps didn't dry, but now they are," Zhyla told us. "Groundwater is going down, and some people's wells in the village nearby have sometimes gone dry—that never happened before."

Vladimir Vernadsky, a Ukrainian biogeochemist, was a "father" of the biosphere concept, as I discussed earlier, and Vasily Dokuchaev, a Russian geographer and soil scientist, had promoted the idea of strict nature preserves, called *zapovedniks*, that would protect areas of intact natural ecosystems as "museum pieces" for scientific study. It therefore felt fitting to be exploring this corner of northwestern Ukraine, where both biosphere reserves and *zapovedniks* protect parts of the unique Polissya ecoregion. Ukraine has eight UNESCO biosphere reserves, one of which, the Shatsk Biosphere Reserve, is not far from the Polisky Nature Preserve we were visiting. Registered in the Man and the Biosphere Programme in 2002, it is part of a tri-national biosphere reserve called the West Polesie Biosphere Reserve, shared by Ukraine, Poland, and Belarus, which was established in 2012. Biosphere reserves are supposed to be laboratories for understanding complex social-ecological systems and models for resolving problems, restoring ecosystem functions and services, and increasing resilience in the face of climate change and other unpredictable events. In Ukraine, the complex of protected areas stretching from the Shatsk Biosphere Reserve to the Polisky Nature Reserve are working on these challenges. But it's complicated, of course.

In our report to the US Agency for International Development's Ukraine Mission, my team and I explained that the forests, wetlands, and bogs of Polissya absorb precipitation, store water, and stabilize river flows between wet and dry seasons. Those ecohydrological processes supply water for domestic consumption, sanitation, irrigated agriculture, hydropower, industry, and transportation, and for environmental flows needed to maintain aquatic species and ecosystems downstream. Because stable flows of clean water depend on biodiverse, functioning, healthy ecosystems, a focus on water automatically provides a link between biodiversity conservation and sustainable development. We recommended that a worthy activity for USAID support, one that would help link biodiversity conservation with economic development in Ukraine, was to "restore wetlands and small rivers in upper watersheds to stabilize downstream flows." We were thinking of Zhyla and his beavers.

~

"Among the agencies which best perform this service of keeping the streams ever-flowing, are the forests and the works of the beaver," wrote Enos Mills in 1909. A century later, wildlife biologist Carol Johnston wrote, "The North American beaver (*Castor canadensis*) is the quintessential ecosystem engineer, causing structural change through its dam building that results in abiotic and biotic environmental changes. The beaver is also a keystone species for riparian obligate animals, providing habitat for many species of waterfowl, wildlife, fish, and invertebrates through its pond building." Putting two and two together, it is not a long leap to the idea that the extermination of beavers in many of the watersheds of salmon country might have had a significant impact on salmon. The overall effect on Pacific salmon of nearly exterminating beavers is not known, but was probably significant. Ben Goldfarb, writing in *High Country News* in 2018, says "Beavers are . . . a keystone species important to the survival of endangered salmon."

It's a bit surprising that such an obvious idea has only started to get traction in Oregon and elsewhere in the Pacific Northwest in the last few years. Way back in 1992, six years before Oregon Coast coho were listed as threatened under the Endangered Species Act, some fish biologists from the Oregon Department of Fish and Wildlife did a study of habitat preferences of juvenile coho salmon in coastal streams. They found that young salmon loved beaver ponds—even though beaver ponds were quite rare then. In their technical, scientific language, they explained:

> During winter, juvenile Coho salmon were most abundant in alcoves and beaver ponds. Because of the apparent strong preference for alcove and beaver pond habitat during winter and the rarity of that habitat in coastal streams, we concluded that if spawning escapement is adequate, the production of wild Coho salmon smolts in most Coho salmon spawning streams on the Oregon Coast is probably limited by the availability of adequate winter habitat.

The obvious message, which was clear more than twenty-five years ago, is that if you want to restore coho, encourage beavers.

Recent scientific support for the benefit of beavers to salmonids comes from a watershed-scale field experiment conducted in north-central Oregon by Nicolaas Bouwes and his colleagues. They constructed 121 "beaver dam analogues"—fake beaver dams—in a watershed that already had beaver activity. Beavers took over most of the BDAs (as they abbreviate them), and also built and maintained 115 natural dams. Monitoring fish populations for seven years showed significant increases in the population of steelhead, a salmonid cousin of coho, as the quantity and complexity of habitats in the watershed increased because of the real and analogue dams. In their study, published in 2016, they conclude, "Beaver mediated restoration may be a viable and efficient strategy to recover ecosystem function of previously incised streams and to increase the production of imperiled fish populations."

In 2016, NOAA Fisheries seems to have finally recognized the beaver-coho connection. In their Oregon Coast coho salmon recovery plan, they noted, "Beaver provide considerable help in providing this connection [between streams and the surrounding forest landscape] and in maintaining proper watershed functioning in Oregon coast streams." They concluded that "Coho salmon recovery demands the application of well-formulated, scientifically sound approaches. The Plan identifies several key steps to improve and protect habitats," including "encouraging formation of beaver dams." The Pacific States Marine Fisheries Commission, mandated by Congress to conserve and manage fisheries in the Pacific Northwest salmon states, is also now touting the benefits of beavers to salmon through its Habitat Program and offering advice about how people and beavers can coexist in the same watersheds.

But sometimes the right hand doesn't know what the left hand is doing. The US Department of Agriculture's Wildlife Services branch has a mission "to provide Federal leadership and expertise to resolve wildlife conflicts to allow people and wildlife to coexist." In Oregon, that has meant trapping and killing "problem" beavers—defined as beavers that build dams that flood farms or private property. Between 2010 and 2016, 3,459 beavers were killed statewide, and 292 were killed in Lincoln County, where the Salmon River is located. So much beaver killing is ironic in a state that has a beaver on the state flag and as the mascot of Oregon State University, has named the beaver the official state animal, and that is informally nicknamed "the Beaver State." In 2017, two nonprofit environmental groups, the Center for Biological Diversity and Western Environmental Law Center, realized that because of the beaver-coho connection, the Endangered Species Act gave them a potential tool to stop the USDA Wildlife Services' beaver killing, so they threatened a lawsuit. Wildlife Services agreed to suspend their beaver control program until they had conducted a study to determine its impact on endangered Oregon Coast coho salmon. Apparently, they are still studying.

Meanwhile, NOAA is funding the installation of beaver dam analogues, following up on the research conducted by Bouwes and his colleagues. (Sergii Mykolayovych in Ukraine will be so excited to hear this news!) In May 2019, NOAA proudly announced that their partners in an innovative pilot project in Oregon "are constructing dam starter structures for beavers to finish building, creating slow water areas for juvenile coho to thrive." One of those NOAA partners, the Wild Salmon Center, a nongovernmental organization dedicated to the conservation of wild salmon that is headquartered in Portland, worked in the upper Nehalem River watershed, not far north of the Salmon River. They reported that "twenty-seven beaver dam analogs were constructed in four high priority reaches for coho, steelhead and cutthroat trout. Beaver found their way in days to the newly built analogs, which are like footholds for the beaver to build out into full-fledged dams using their advanced engineering skills."

~

Beavers build dams, creating disturbances and resetting succession along streams in a mosaic that shifts in both space and time. With beavers, watersheds experience what beaver ecologist Robert Naiman and colleagues describe as "multisuccessional pathways." In trying to understand and explain how an ecosystem engineer and keystone species like *Castor canadensis* fits into our views of ecological change over time, the authors' bottom line was, essentially, "it's complicated." They describe how beavers interact with landscapes by saying that "factors responsible for individual successional pathways include existing vegetation, hydrology, topography, fire, disease, herbivory, and beaver. The multisuccessional pathways are complex." Figures in the 1988 article by Naiman and colleagues, "Alteration of North American Streams by Beaver," show looping flow diagrams, spaghetti charts of interconnections, ecological spiderwebs.

And the very complexity of those interconnections has caught the attention of other ecological theorists, who have suggested that "streams with beaver ponds probably have a high resistance to disturbance" and that "beaver also assist in returning the stream to a predisturbance condition." So, beavers are also, it seems, architects of ecological resilience—the resistance to ecological disturbance and ability to return to predisturbance conditions.

"Beaver landscapes are cauldrons of turmoil and bastions of stability, dynamic and durable in equal measure," says Ben Goldfarb in his book *Eager*. Maybe there is a message here for us humans. In a chaotic world, interconnections can create systems that can absorb dramatic, dynamic change but still regain equilibrium. Maybe a message we can learn from beavers.

~

At Cascade Head, eager beavers have been busy, quietly assisting in the restoration of watersheds and salmon, and not only at Pixieland. At about the same time that Pixieland was being developed in the late 1960s, a trailer park called Tamara Quays was constructed on filled marshland along Rowdy Creek. The site was restored from 2007 to 2009 by removing concrete trailer pads, the septic systems, fill, and a tide gate, and reconnecting the flow of the creek and ensuring fish passage with new culverts under a nearby road. Before long, juvenile coho were seen jumping over a beaver dam above the restoration site to gain access to habitat higher upstream. Nearby, a tributary of Rowdy Creek drops from the ridge to the south through land owned by Camp Westwind and crosses the Y Marsh, the section of salt marsh reconnected to tidal flows in 1987. Perched near the top of the short watershed, which is only around a mile and a half long, is Lost Lake, a small pond created by a beaver dam that plugged the creek. According to Duncan Berry, a longtime Cascade Head resident, it had been there a while when he first saw it in 1971, and beavers were active there until maybe six or seven years ago.

On the other side of the Salmon, at Crowley Creek, a dike was removed and a new, larger culvert was installed under Three Rocks Road in the summer of 2012 to reconnect the small creek to the estuary. During the restoration work, beavers showed up, perhaps attracted by the splash of water through the culvert, and cut several large trees each night. Eventually they built an impressive dam across Crowley Creek, and the colony was active for years.

When I was out on the Salmon River in the kayak looking for the layer of sand and mud from the 1700 tsunami in the marsh sediments, I found a piece of beaver-cut wood on the river bank. It was a nine-inch-long section of a three-inch-diameter alder trunk, whittled sharply with neat rows of tooth marks to points at both ends, and with the bark gnawed off in a spiral pattern to reveal the white wood. Just a small chunk of wood, like a little football. I wondered why a beaver would go to that much trouble, gnawing through that much wood twice, to make such a small piece. My scientific hypothesis? I think it was a beaver version of a message in a bottle, sent downstream to say, "Hello! We're here!"

To say, perhaps, we're *still* here, despite what you did to us—resilient in this dynamic landscape that we shape and make still more dynamic, as we have done across the northern continents since our ancestors were giants. We'll still be here awhile longer. And you?

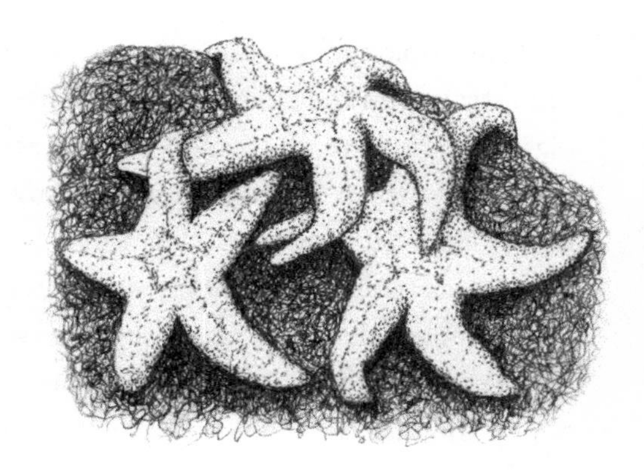

11 *Where Have All the Seastars Gone?*

When to the new eyes of thee
All things by immortal power,
Near or far,
Hiddenly
To each other linkèd are,
That thou canst not stir a flower
Without troubling of a star . . .

—Francis Thompson, from *The Mistress of Vision,* 1897

The beach north of Roads End ended in a black cliff of breccia. A wave-cut terrace extended seaward around the western base of this point, toward wave-washed Polly Rock offshore. Swell surged through deep channels, and bands of California mussels (*Mytilus californianus*) covered the slick black basalt starting at about eye level from where I waded to my ankles around rocks and along channel walls. The mussel bands were only about a meter wide, often topped by colonies of white-shelled goose barnacles. On this overcast, low-tide, late afternoon in early January 2019, everything was black and white and silver-gray, like an old platinum print, and the place had an edgy, dangerous feel, especially after an unexpectedly high wave surged up a channel where I was wading and wet me to my waist. I didn't want a swim in that cold ocean.

I was looking for seastars. I searched the nearby rocks below the band of mussels, and the sides of unreachable channels farther out with binoculars. Where have all the seastars gone? This wasn't normal, in my experience. I've spent probably a year of my life in the intertidal zone of the Pacific Coast, if I added it all up, and I always expect to see constellations of seastars, as much as I expect to see stars in a clear night sky. Finally, across a channel too deep to cross, I spotted two orange ones; with binoculars I could make out a few of their brown and purple siblings nearby. Orange is one of the several color variants of this species, from which it gets its scientific name *Pisaster ochraceus*, the ochre seastar. I counted only a handful, but they appeared to be good-sized adults, which I took as a positive sign. I clambered over volcanic dykes and boulders to reach a cove around the point to the north. From one vantage, I saw a wave-swept vertical face, again unreachable at this tide, with a few more seastars, including an orange one with only four legs, the base of the fifth a whitish stub—an indication that sea star wasting syndrome is still lurking here. The dramatic pinnacle of rock called The Thumb towered above the cove, and the view north looked out between black spires to the western cliffs of Cascade Head. On the way back around Roads End Point I finally found a *Pisaster* close enough to touch, a medium-small purple one about four inches across. Maybe one of the post–wasting generation, I thought.

I'd seen fewer than a dozen *Pisaster* all afternoon, in this area where there should have been hundreds. But the mussels were still in a tight, narrow band, apparently not spreading out to take over habitat lower in the tide zone. The bare, wet rock below them seemed bizarrely barren though—no mussels, no barnacles, no clumps of algae. What's going on here?

My fascination with seastars started as early as I can remember. On summer visits to Oregon, my grandfather took me to the tide pools of Haystack Rock, at Cannon Beach, where seastars crammed in every crevice and crawled on every rock. I remember a special delight in their diverse colors—brown, orange, and some a deep, royal purple, like the color of the robes of the ancient King of the Sea. My granddad had a technique for boiling, salting, and drying a few seastars we'd collect to send home with me as souvenirs of our time together. When boiled, they all turned orange, like cooked crabs or shrimp do. We set them to dry in the sun, poured salt on their undersides, and scraped the shriveled tube feet out of the grooves on their arms with a knife when they were dry. For the trip home, my dad sometimes tied them to the bumper because they still smelled too strong to take into the car, or even put in the trunk. In dry New Mexico they mummified quickly, and for the rest of the year lined my bedroom windowsill, pentamerous radial suns shining their magical memories on me until the next summer. A few would disappear now and then during the year if Blackie, our black Labrador, snuck in and snitched one for a snack. After I got a PhD in ecology, I remember blurting out, when someone asked me, "Well, how did you come to be an ecologist?" that it was all because of my granddad, and the tide pools at Haystack Rock. Later I wondered, Did I really mean that—that I became an ecologist because of those experiences starting at five years old? And when I thought harder and deeper about it, all I could come to was—Yes!

~

Chapter 6 of *Cannery Row*, John Steinbeck's fond, short novel about his marine biologist friend Ed Ricketts, starts like this:

> Doc was collecting marine animals in the Great Tide Pool on the tip of the Peninsula. It is a fabulous place: when the tide is in, a wave-churned basin, creamy with foam, whipped by the combers that roll in from the whistling buoy on the reef. But when the tide goes out the little water world becomes quiet and lovely. The sea is very clear and the bottom becomes fantastic with hurrying, fighting, feeding, breeding animals.

Steinbeck proceeds to describe in colorful detail some of the hurrying, fighting, feeding, and breeding, and makes the Great Tide Pool into a microcosm and metaphor for all of the processes of life and evolution. *Cannery Row* was published in January of 1945, before the Third Reich was defeated in Europe, before the world learned of the Manhattan Project and the atomic bomb, and it reflects something of the "hurrying, fighting" human condition of the time.

In his ecologically oriented field guide to the Pacific intertidal zone, *Between Pacific Tides,* Ricketts described ecological communities and their physical environments, rather than working taxonomically from flatworms to chordates as academic marine biologists had done previously. Stanford University Press finally published *Between Pacific Tides* in 1939, after some opposition from academic reviewers. But Ricketts's approach provided a guide to the Pacific shore for curious, nonacademic, lay naturalists, and the book eventually became one of the top-selling titles in the Stanford University Press catalogue. It was the book I got through interlibrary loan from the Los Alamos Public Library before going to Camp Arago, a marine science camp on the Oregon Coast, when I was in high school. It was the book that got me hooked on intertidal ecology.

The wonder of the seastar-filled tide pools at Cannon Beach led me eventually—somehow inevitably, I now suppose—to my PhD research on the ecology, behavior, and genetics of a common intertidal snail of the North American Pacific Coast, the black turban snail, whose scientific name is *Tegula funebralis*. I'd become acquainted with *Tegula* at Camp Arago and read about it in *Between Pacific Tides*. As I was exploring sites for my thesis research, I thought it would be very special if Doc's Great Tide Pool was one of them. Getting permission to mess around in the Great Tide Pool took some doing, but I was aided and abetted by a faculty friend at Stanford University's Hopkins Marine Station, Chuck Baxter, who had taught my undergraduate marine biology class. I considered it quite a coup. But Doc's Great Tide Pool turned out not to have all the "right stuff" for my research, so I did my first experiments on habitat choice in black turban snails at another tide pool I knew from my days on the staff at Camp Arago. That tide pool, at Middle Cove on Cape Arago, near the fishing port of Charleston on Coos Bay, became my own Great Tide Pool.

The pool was a meter or so deep (about a yard) at low tide, and above it an angled slab of sandstone rose steeply landward, covered with bands of algae that reflected the dramatic ecological change, over a very short distance, from sea to land. Black turban snails were abundant, both in the pool and up the sloping wall above it for a couple of meters. It was an ideal place to ask the snails a question: Do you care where you are, in the pool or above it? Do you choose your habitat?

How do you ask a snail that question? That first exploratory summer of research, I collected a few handfuls of snails from the tide pool, and a few more from the rocks above it. I dried their shells, and dabbed them with a splotch of paint—blue for "tide pool" and yellow for "rock"—and then switched their habitats. I put the blue-marked snails on the rocks, and the yellow-marked ones in the pool. Over the next few days, the snails gave me the answer: We *do* care, we *do* choose! Most of the blue snails crawled back into the pool, and most of the yellow ones crawled back up onto the rocks. Back on campus in September, my doctoral thesis adviser said, "Go for it," and so I spent the next two summers chasing black turban snails at several sites up and down the Pacific Coast, from La Jolla, California, to Bamfield, on the central coast of Vancouver Island, British Columbia.

Pisaster ochraceus, the ochre seastar, is a common community member of the open coast rocky shore ecosystems of the Pacific Coast. "Since these animals are not 'fish' in any sense, 'seastar' is perhaps a better name for them than starfish, and has been used throughout this book," Ricketts wrote in *Between Pacific Tides*. Indeed, they are not vertebrates like fish, but echinoderms, the "spiny-skinned" creatures of the phylum Echinodermata, a major branch of the evolutionary tree of animals that includes seastars, sea urchins, sand dollars and sea cucumbers. The members of this diverse, ancient group of creatures share several characteristics: spiny skins, suction-like "tube feet" with which they attach to the substrate, and, generally, "pentamerous radial symmetry." At least for the seastars, that last characteristic means they are often shaped like five-pointed stars.

Ochre seastars are fierce but slow predators, the most important carnivore of the intertidal zone. Ricketts explained that "*Pisaster* will eat almost anything it can get its stomach onto, or around; the stomach is everted and thrust into the shell of a clam or mussel, or around a chiton or snail. Digestion, therefore, takes place outside the body of the diner, but the stomach must be in contact with the meal." He described how a seastar can fasten onto the shell of a California mussel with its tube feet, pull the shell apart just enough to insert its stomach through the crack, and digest it cleanly inside its own shell. Mussels are the preferred and most important food of this seastar throughout most of their range, but in habitats where mussels are scarce, they can shift to barnacles, chitons, abalone, or snails, including black turban snails. The first systematic study of their diet was reported in a 1959 paper by Howard Feder, who pried 3,450 *Pisaster* off the rocks at low tide to see whether, and what, they were eating. He found they were eating something 18 percent of the time, and noted thirty-three different prey species. The most common prey, numerically, were small barnacles, which made up more than 50 percent of the observed prey items. Mussels were next, accounting for 17 percent of seastar meals; but because mussels are so much bigger than barnacles or other common prey, they made up by far the largest part of the *Pisaster* diet in terms of calories, if not in numbers eaten. In Feder's study, the black turban snail was the fourth-most-common prey species eaten by *Pisaster ochraceus*, making up about 6 percent of all prey items.

In graduate school I'd become familiar with the fascinating research on *Pisaster* and its feeding ecology being done by Robert Paine, a professor at the University of Washington. Paine had started working on the outer coast of the Olympic Peninsula at Mukkaw Bay, Washington, in 1963. (This is now spelled Makah Bay to conform to the spelling of the name of the Makah Tribe; I'll use the old spelling, as Paine did, when referring to his research.) He removed all the seastars from a stretch of rocky intertidal shore eight meters long and watched what happened over the next few years. An adjacent area with seastars served as the experimental control. Before removal of seastars, California mussels and goose barnacles formed a thick band across the middle of the intertidal zone, below which various grazers like limpets, chitons, and snails cruised among clumps of attached algae. When seastars were removed, acorn barnacles began to colonize the rocks below the mussel beds, and before long mussels and goose barnacles spread downward; the algae were crowded out, and the grazing gastropods left for greener pastures. Removing *Pisaster* resulted in simplification of the food web in the area and a decrease in species diversity, from a system supporting fifteen species to one with only eight species, Paine found.

By 1966, Paine had published a landmark paper on this experiment in *The American Naturalist*, titled "Food Web Complexity and Species Diversity." It described how *Pisaster*, as the top predator in the intertidal food web, played a key role in controlling the number of species present in the ecological community. By eating lots of mussels, seastars stop them from capturing most of the space on intertidal rocks and thereby limiting the real estate available to many other species. Paine followed up on that research in a 1969 paper, in which he first used the term "keystone species" for dominant predators like *Pisaster*. The term, and the concept, quickly took hold among ecologists and has shaped thinking about community ecology, biodiversity, and conservation ever since.

Another of Paine's papers, in the journal *Ecology* in 1969, was called "The *Pisaster-Tegula* Interaction: Prey Patches, Predator Food Preference and Inter-Tidal Community Structure." It was also based on his work at Mukkaw Bay. Here's a bit of the abstract, to give the flavor:

> The herbivorous gastropod *Tegula funebralis* is not highly ranked in a food preference hierarchy of its major predator, the starfish *Pisaster ochraceus,* and exhibits a persistent broad overlap with it in the rocky intertidal zone at Mukkaw Bay, Washington. *Pisaster* consumes 25–28% of the adult *Tegula* per year in the area of spatial overlap. . . . It is suggested that the implied results of the interaction is typical of that between a major predator and one of its less preferred prey. The prominent zonation exhibited by preferred prey, [and] the observed intimacy of association of predator and less preferred prey . . . are discussed.

I knew if I was going to study *Tegula,* I would have to pay attention to *Pisaster*—and to Paine.

I had already read his paper on the relationship between ochre seastars and black turban snails before my first summer of habitat-choice experiments and was well aware that part of the question I was posing to the snails was, "where are the seastars?" Imagine a black turban snail's reaction to Paine's description of them as "not highly ranked in a food preference hierarchy" of the seastar, and "one of its less preferred prey." If you were a *Tegula,* you would be saying "Holy shit!—they are eating a quarter of us every year, and that crazy so-called professor calls us a 'less preferred prey'!"

I remembered a passage in *Between Pacific Tides,* where Ricketts says, in his idiosyncratic, poetic style (which I love!),

> Some years ago E. C. Haderlie observed that limpets would move away from a seastar on a rock as rapidly as possible; they were thrown into a sort of hysteria, as he put it, and for a limpet, the speed was virtually a gallop. Many snails (including abalone as it turns out) will dissociate themselves from the presence of *Pisaster*, usually with as much alacrity as they can muster. It is not necessary for them to be touched by the seastar; its nearby presence is enough to set them on the move.

Based on that description, I got an idea for how to test the reaction of *Tegula* to the presence of *Pisaster* nearby. I used a pair of one-liter graduated cylinders—a common piece of laboratory glassware; they are a glass tube about a foot tall and two inches in diameter. To start the experiment, I would fill each cylinder half-full of seawater and put a dozen black turban snails in each one. Then I would fill one cylinder to the top with more plain seawater and the other with seawater from a bucket that held an ochre star. The result was instant, entertaining, and definitive. The snails in the tube topped up with more plain seawater were their usual lethargic selves; they hardly moved a muscle. But as soon as the *Pisaster*-water went into the experimental cylinder, the snails started thrashing and climbing up the sides. Ricketts's description that "they were thrown into a sort of hysteria" and fled upward at "virtually a gallop" with "as much alacrity as they can muster" was a perfect description of the observed behavior. There was no question that black turban snails know about, and pay attention to, the presence of the most fearsome keystone predator in their cosmos.

My initial experiments with the blue- or yellow-marked turban snails at Cape Arago showed that most snails found either above or in permanent tide pools at low tide return to their original habitats within a few days after experimental habitat reversal. The next summer field season, wanting to confirm the generality of my results at another location, I immediately thought of Mukkaw Bay. Paine had done his famous *Pisaster* removal study at a mainland site (he later moved his experiments to Tatoosh Island), and I wanted to work there if I could find a suitable tide pool. I imagined that piggybacking on Paine's research might give more insights into the behavior of black turban snails. I visited his lab in Seattle, told him about my research, and before long we were looking at maps and he was giving his blessing for me to work at his site at Mukkaw Bay. When I went there, I found a suitable tide pool, not quite as deep as at Cape Arago, and a little higher in the tide zone, and it became my Great Tide Pool No. 2.

Mass-marking studies of habitat choice like I'd done the first year can't detect more detailed preferences, so in the second year I collected and marked individual snails so I could keep track of their exact position in the intertidal zone. I marked hundreds of snails, at both Cape Arago and Mukkaw Bay, with tiny tags that professional beekeepers use to mark honeybee queens—plastic discs about the size of a capital "O" on this page that came in five colors, each with a number from one to ninety-nine. I would first measure the original position of a snail above or below the tide pool surface, glue a tag to its shell with Superglue, then place it and all the other snails I'd marked at the pool surface every morning and remeasure their positions the next day. During a two-week low-tide series, I got information about thousands of individual habitat choices.

My research showed that individual snails returned to approximately the same intertidal level where they were originally found; their habitat preferences were more nuanced than merely for above-pool or in-pool habitats. The intertidal position of snails was influenced by age, the presence of seastars nearby, and intertidal temperature contrasts at low tide. Following warm, sunny days, when temperature differences between in-pool and above-pool habitats are large, snail populations move lower in the intertidal zone and constrict their range compared with that on days following cool, overcast weather. I also found genetic differences associated with habitat choice that seemed to be associated with temperature tolerance. It appeared that individual snails were balancing various factors in their habitat-choice behavior. Think about it: If you were a black turban snail, wouldn't you want to live in a nice, comfortable, even-temperature tide pool, where your gills never dried out and you were never rinsed with horrible-tasting fresh rainwater? I can imagine at least two reasons why not: the risk of being eaten by a seastar in the pool, and the prospect of better grazing on the out-of-water, less-grazed rocks above.

For my PhD thesis research, I "sacrificed" hundreds of black turban snails from my research tide pools, ground them up, kept their juices in test tubes in a minus-70-degree Celsius freezer, and examined their genetic makeup using the best techniques of the day. We actually used the word "sacrificed," because as ecologists, we loved and empathized with our research subjects, but we were confident that the gods of scientific knowledge would forgive our sins. In another worldview, maybe my karma will see me reborn as a black turban snail, and I'll understand. Come to think of it, for all those seastars I boiled and salted with my Granddad Sweeney, I'll probably come back as an ochre star too, for at least a few lifetimes. Oh well, in either case I don't think it would be so bad. In one reincarnation I'd be the chased and eaten, in the next the chaser and eater.

In the end, I had an answer to the question, "Is behavior ecologically adaptive?" In black turban snails, the answer is "yes." My doctoral research led on to other habitat-choice studies, like that on the ribbed limpet I described earlier, where the answer to the question was also a definitive "yes." And in black oystercatcher feeding traditions, a clear "yes." I'm still debating whether, in my own species, "yes" or "no" predominates. But I'm convinced that in some of our behaviors, the answer is yes. What we choose does have ecological effects; our choices will affect our future, maybe even our survival. Our choices do matter.

~

The ochre seastar, *Pisaster ochraceus*, has helped ecologists learn some important lessons about the structure and stability of ecological communities, but in a number of ways it is also still teaching us how much we don't know. In his study of *Pisaster* diets, published in 1959, Howard Feder wrote, "Although the star-fish *Pisaster ochraceus* is one of the most conspicuous animals to be found along the rocky shores of the Pacific Coast, its natural history is poorly known." Some recent research shows that is still true today.

For example, *Pisaster*'s orange-brown-purple color polymorphism that had so delighted me at age five is still a mystery. A 2007 paper that used a large-scale biogeographic approach to try to explain it began by confessing that "remarkably little is known about the basic biological foundations of its color variants nor the ecological conditions that support phenotypic color polymorphism in this seastar." Peter Raimondi and his coauthors reviewed the quite-extensive ecological literature on color variation, which showed that in the many species in which it is found, coloration is ecologically adaptive and readily explained by evolutionary theory. They stated that general conclusion in "scientese," saying, "In almost all studies, the null hypothesis that color polymorphism is selectively neutral has been rejected." I've already described my research on shell color diversity in the ribbed limpet, which allows brown limpets to be camouflaged on rocks and white ones to hide among goose barnacles. I also studied color polymorphism in another snail, the rough winkle, *Littorina saxatilis,* on the coast of Wales, and showed that its rainbow of shell colors helped it hide from visually hunting predators on rocks ranging in color from red sandstone to gray limestone. So, I have always wondered about an ecological and evolutionary explanation for the color variation in *Pisaster,* and guessed that there is one just waiting to be discovered.

But Raimondi and his colleagues didn't really find an answer. They compiled thousands of observations of seastar color and size from southern California to northern Oregon. Throughout that range, they found that orange seastars—the "ochre" variant reflected in the scientific name *ochraceus*—usually make up approximately 20 percent of the population, and the frequency of orange increases with the size of individuals in most populations. Their non-conclusion conclusion was, "These novel findings point to the need for renewed study of the basic biology of this key ecological species."

Another study of *Pisaster* color, by Christopher Harley—who was one of Bob Paine's graduate students—and his colleagues, extended the observation past the Oregon border, up the coasts of Washington and Vancouver Island and into the Strait of Juan de Fuca, Puget Sound, and the Salish Sea. They reported that while *Pisaster* populations along the outer Pacific Coast are mostly reddish-brown or dull purple, from 6 to 28 percent are orange, and a small percentage are brilliant purple. But if you turn the corner at Cape Flattery, into the quieter waters of the San Juan Islands and the southern Strait of Georgia, almost all are bright purple. They wondered whether the dramatic *Pisaster* color polymorphism had some genetic basis and used an analysis of mitochondrial DNA to look at the genetic structure of *Pisaster* populations from central California to Alaska. But they found no significant genetic differences across that vast range, suggesting a high level of genetic mixing of *Pisaster* populations. That makes sense, because these seastars spawn by dumping their eggs and sperm into the ocean, and coastal currents could easily carry the offspring of spawning *Pisaster* from Alaska to Oregon, or from British Columbia to California. Even so, this study could not rule out a genetic association with color: "In some cases, color morphs are genetically distinct. . . . However, in other cases, there may be little association between coloration and genetic structure." Their conclusion was that "the factors that maintain the color polymorphism, and those that contribute to among-site variation in color frequencies, remain unknown."

When a future PhD dissertation finally explains the color variation in *Pisaster,* I'm sure it won't involve an explanation as simple as in my studies of the phenomenon—hiding from visually hunting predators through a diversity of colors that provide camouflage on different backgrounds. As Ed Ricketts explained in *Between Pacific Tides,* the intertidal bible, "*Pisaster* neither has nor seems to need protective coloration. Anything that can damage this thoroughly tough animal, short of the 'acts of God' referred to in insurance policies, deserves respectful mention."

~

Respectful mention? Acts of God? Insurance policies? Beginning in June 2013, seastars on the San Francisco Peninsula of California and Olympic Peninsula in Washington were seen with the following gruesome symptoms: lesions, tissue deterioration, loss of rigidity, twisted arms, arms falling off, and internal organs emerging from lesions. This mysterious seastar disease led to loss of hydrostatic pressure in the water vascular system that creates suction in the tube feet, so they lost their grip on the rock. They were literally melting, falling apart. Sea star wasting disease (SSWD) or sea star wasting syndrome (SSWS), as it came to be called, first appeared in Oregon in April 2014, and by June had spread to most of the Oregon Coast. In less than a year, seastars with these symptoms were found along virtually the entire Pacific Coast. By 2015, it had spread from Baja California to Alaska, with huge declines in *Pisaster ochraceus* populations. The disease affected twenty-two species of seastars to varying degrees, some of which, like the ochre seastar, are keystone predators. Among the other species hit hard by SSWS were the sunflower star, *Pycnopodia helianthoides,* the sun star, *Solaster dawsoni,* and the mottled star, *Evasterias troschelii.*

When sea star wasting struck in 2013, two large multi-partner marine monitoring networks were already in place, and they sprang into action to study this unprecedented phenomenon with its potentially huge ecological consequences. Both networks had long-term data on seastars and intertidal communities at study sites along the Pacific Coast reaching back twenty-five years before the SSWS outbreak, providing a baseline against which its ecological effects could be measured. Although limited episodes of seastar wasting had been observed before, nothing on this scale had been seen. It was one of the largest outbreaks of disease in a marine ecosystem ever observed, and surely one of the best studied.

In some ways these monitoring networks were a legacy of the academic "family" of Robert Paine, often involving researchers who had been his students. One of the networks, PISCO, the Partnership for Interdisciplinary Studies of Coastal Oceans, was founded in 1999 by a couple of Paine's graduate students, Bruce Menge and Jane Lubchenco, now professors at Oregon State University. It links a number of marine laboratories to conduct collaborative studies of the California Current Large Marine Ecosystem, a 1,200-mile region of the Pacific Coast stretching from Baja California to Vancouver Island. PISCO combines "experimental research questions with long-term monitoring across sometimes large spatial scales to examine causes and consequences of ecological changes relevant to species and their conservation and management," according to its website. The other large network is the Multi-Agency Rocky Intertidal Network (MARINe), a consortium of research groups that have used standardized protocols to collect comparable ecological information at more than two hundred rocky intertidal sites, some of which have been studied for thirty years. MARINe has fourteen monitoring sites in Oregon. Roads End Point, where I'd searched for seastars that January day, is one of them.

All of this monitoring showed that *Pisaster* numbers decreased by from 54 to 97 percent across its range, and seastar biomass dropped to 10 to 20 percent of baseline levels. Predation rates on mussels were dramatically reduced at most sites, and, as predicted by Paine's pioneering experiments, the lower edge of mussel beds shifted downward in many places. But the seastar die-off seemed to trigger a baby boom of sorts, surprising biologists; in 2015, tiny *Pisaster* settled in unprecedented numbers at many sites almost barren of adults—up to three hundred times more little seastars than normally seen. (Ah, Walt Whitman! Always the procreant urge of the world, I suppose.)

Roads End Point lies within the core area of the Cascade Head Marine Reserve, and thus within the core protected-area zone of the Cascade Head Biosphere Reserve. No harming or harvesting of invertebrates, seaweed, fish, or other wildlife is allowed. The shoreline and offshore marine environment in the Marine Reserve are meant to be another of those ecological reference areas, a marine "museum piece" or *zapovednik*. The Nature Conservancy has been monitoring seastars at Roads End Point since October 2014 as part of the MARINe network. Using the network's standardized protocol, Dick Vander Schaaf, TNC's associate coast and marine conservation director for Oregon, has been counting and measuring *Pisaster* every few months. As was observed elsewhere up and down the coast, in October 2014 there were very few *Pisaster* left at Roads End Point, and numbers stayed low until early 2017, when they began to climb steadily. The average size of individuals also started to increase. Those trends have held until the last census at Roads End, in August 2018, and it appears that *Pisaster* are coming back. But if so, they have a long way to go to reach the densities that I was expecting to see on my low-tide visit in January 2019.

Within a year after the SSWS epidemic began, a research team identified a possible cause as a pathogen called a "sea star-associated densovirus." But that finding didn't really explain what triggered the outbreak, because the densovirus was also present in seastars without symptoms of wasting, and exposing seastars to the virus didn't always cause wasting. Perhaps some environmental change triggered it?

The latest research, by Ian Hewson and his colleagues in 2018, basically concludes that this dramatic phenomenon is a scientific mystery. Viral pathogens don't entirely explain it, and high water temperature or low precipitation don't explain it. "We speculate that SSWD may represent a syndrome of heterogeneous etiologies [i.e., diverse causes] between geographic locations, between species, or even within a species between locations."

Melissa Miner of the University of California, Santa Cruz, led a team of more than a dozen researchers who tapped the MARINe database to look at the large-scale impacts of the seastar wasting epidemic and the implications for ecological recovery. In their 2018 report they say, "In hindsight, our data suggest that the SSWD event defied prediction based on two factors found to be important in other marine disease events, sea water temperature and population density." As far as the recovery of seastar populations to former levels, they say that "low levels of SSWD-symptomatic sea stars are still present throughout the impacted range, thus the outlook for population recovery is uncertain."

Can an intertidal keystone species recover naturally from a mystery disease? Observations at Roads End Point, and up and down the coast, seem to say a hopeful "yes." Seastars and their world seem to have a kind of resilience, even if we don't understand why.

~

While looking into sea star wasting syndrome, I learned about a creative collaboration among seastar researchers, conservationists, and the private sector from Jenna Sullivan, a PhD student in Bruce Menge's lab at Oregon State, and one of his star seastar researchers. I guess you could say she is a granddaughter in Bob Paine's academic "family." In a moment of inspiration, she wrote an email to a contact at Rogue Ales and Spirits, whose production facility sits just down the road from OSU's Hatfield Marine Science Center in Newport, and proposed that perhaps they would like to support work on SSWS. "They had been involved in conservation-type stuff before, so I thought it would potentially be a good fit," Jenna told me. Rogue's response was to brew a special beer, dedicated to the suffering seastars, called Wasted Sea Star Purple Pale Ale, and give a portion of the proceeds from its sale to PISCO's SSWS research. The brewmaster used purple corn nectar, made from blue corn, to give the ale a purplish hue, playing on the common name "purple seastar," which many people in Oregon use instead of "ochre seastar." Beer, science, and conservation seem to go together, at least in Oregon. In 2012, Pelican Brewery of Pacific City, Oregon, released a special brew called Silverspot IPA. "Drink a beer, save a butterfly!" they proclaimed, giving part of the profits to organizations working to conserve the Oregon silverspot butterfly.

With our pints of purple Wasted Sea Star Purple Pale Ale in hand, let's circle back in time and propose a toast to the cascade of ecological ideas initiated by the research collaboration between *Pisaster ochraceus* and Robert Paine. ¡*Salud*! Cheers!

Paine's research focused on what was going on in specific, local places, like the tide pools of Mukkaw Bay, but ended up discovering broadly applicable concepts and general principles. Now, according to Bruce Menge, "The problems we're trying to solve in ecology are way beyond the local scale."

Seastar wasting syndrome has everyone worried about, and scientifically fascinated with, the stability and resilience of intertidal ecosystems, and ecosystems in general. I've been making the claim throughout these essays that resistance to the forces that would destroy them can protect intact natural ecosystems, or at least examples of those, so that research can understand how they function; and, based on that understanding, we can restore them; and then we will move toward the human-nature reconciliation, which will lead to ecological resilience. That's the logical progression of the five "re"s that have framed my narrative so far. But in this story of keystone seastars and sea star wasting syndrome, there is a link missing. We are left struggling to do the research to understand how ecosystems function, while we helplessly watch them change, and—perhaps—restore themselves with no help from us, through their natural resilience. All we can do is watch, worry, wonder, and trust nature to work it out.

Ecology is the humbling science. A lesson that we should have learned from the ochre seastar, *Pisaster ochraceus*, is that we don't know, and likely can never know, all of the mysteries of ecological systems, even though, like them, we are a keystone species. Perhaps, therefore, it is fitting to end with a small sermon from *Log from the Sea of Cortez*, the little gem of a book describing a six-week marine intertidal collecting expedition John Steinbeck and Ed Ricketts made from Monterey, California, around the Baja California Peninsula and into the Gulf of California in the spring of 1940:

> It is a strange thing that most of the feeling we call religious, most of the mystical outcrying which is one of the most prized and used and desired reactions of our species, is really the understanding and the attempt to say that man is related to the whole thing, related inextricably to all reality, known and unknowable. This is a simple thing to say, but the profound feeling of it made a Jesus, a St. Augustine, a St. Francis, a Roger Bacon, a Charles Darwin, and an Einstein. Each of them in his own tempo and with his own voice

> discovered and reaffirmed with astonishment the knowledge that all things are one thing and that one thing is all things—plankton, a shimmering phosphorescence on the sea and the spinning planets and an expanding universe, all bound together by the elastic string of time. It is advisable to look from the tide pool to the stars and then back to the tide pool again.

Amen!

12 *Whale Haven*

Y Dios creó las grandes ballenas
allá en Laguna San Ignacio,
y cada criatura que se mueve
en los muslos sombreados del agua.

—Homero Aridjis, *El ojo de la ballena,* 1999

It was a pure blue October day, not much wind, but the ocean was restless, the deep, slow pulse of a faraway Pacific storm surging and sucking on the headland north of Rocky Creek. And there were whales, blowing and diving inside the outer line of foam. I saw two immediately. When they surfaced after a feeding dive, they would blow, and blow, and blow, and blow; four, five, or six times, mostly lolling in one place. Then, after the last breath, their backs would arch up, the dorsal hump and row of bumps along the spine behind it would slide under the surface, and sometimes the wide flukes would show as the whale headed down for another bout of feeding. In four or five minutes, they would surface again to breathe, and then after another series of explosive breaths, dive again. It was hard to keep track of how many whales were out there; as one seemed to drift south along the coast, another would appear from the north. The whales didn't fluke up on every dive, and even when they did, the orientation had to be just right for a clear view, but I saw at least two distinct patterns. One whale's flukes were clean and black. The other had what I called "paint drips" of white—as if someone had painted the outer edges with white paint, and it had run down in trickles, leaving a drip at the end. I didn't know it then, but later learned that "Paint Drip" carried the scars of an encounter with orcas somewhere along her life's journey—not unusual for gray whales.

I watched for what seemed a long time; I'm not quite sure how long it was really. Land time was suspended, and rhythmic slow-motion sea time lulled me into a deep whale meditation. Before long, I knew these locally feeding whales would stop feeding here and join the parade of their kin from the north, heading south for the lagoons of western Baja California, where all gray whales in the eastern Pacific give birth. When I finally looked at my watch, I realized I had to hurry on south to Newport or I would be late for a meeting there.

~

The gray whale, *Eschrichtius robustus*, is a baleen whale that migrates along the Pacific Coast of North America from calving grounds in Mexico to feeding grounds in the Arctic. These whales travel five thousand to six thousand miles in each direction, moving across 50 degrees of latitude, one of the longest annual migrations of any mammal species. During summer and fall, most whales in the eastern Pacific population feed in the Chukchi, Beaufort and northwestern Bering Seas between Alaska and Siberia. Most spend the winter along the west coast of Baja California, where breeding and birthing take place. This species used to inhabit the Atlantic Ocean as well, migrating up and down both the North American and European coasts—but those populations were driven to extinction from whaling by around 1750. An endangered remnant population of gray whales is hanging on in the western North Pacific around Sakhalin Island, the Kamchatka Peninsula, and in the Sea of Okhotsk. A 2016 report estimated this western Pacific population at between three hundred and four hundred adults. A few satellite-tagged whales from this population have been spotted in Mexico, and no breeding grounds have been identified in the western Pacific. Where these whales breed and calve is a scientific mystery that remains to be solved.

Gray whales grow to between forty and fifty feet long, and weigh forty to forty-five tons (about ten times the weight of an average SUV or pickup truck). They can live seventy years, not much different than humans. Gray whales are specialized for feeding in shallow coastal waters, distinguishing them from their baleen-whale cousins—blues, humpbacks, minkes, and others. All baleen whales feed on small crustaceans and fish, which they strain from the water using a rack of thin, closely spaced plates called baleen that extend from their upper jaw. Baleen is made of a flexible, fibrous protein called keratin, the same material that forms mammalian hair, nails, claws, hooves and horns. Gray whales have about three hundred of these baleen plates, each around a foot long, with bristles on the inside. They feed by sucking in a gulp of water containing small crustaceans or other prey, then forcing the water out through the baleen sieve with their tongue, concentrating the food so they can swallow it. Grays are versatile, opportunistic feeders and can harvest prey from bottom sediments and also from open water off the bottom. On the summer feeding grounds in the Arctic, their diet consists mainly of amphipods—tiny shrimp-like crustaceans that they suck off the seafloor in relatively shallow water. Gray whales are the only whale known to feed on prey in bottom sediments in this way. Whales feeding in the summer and fall along the Oregon Coast, like the ones I was watching at Rocky Creek, seem to eat a lot of mysids, commonly known as opossum shrimp, which can be abundant in and around kelp beds. Crab larvae, worms, and small fish can also be part of their diet. During months of summer feeding, whales put on a season's worth of blubber before heading south in late fall. They won't feed much again until the next summer.

Most gray whales have left the northern feeding areas by late November. The southbound whale parade is led by pregnant females, eager to get to the calving lagoons. Then come mature males and females ready to mate that year, often mating as they travel south. Immature whales of both sexes straggle along at the back of the pack. During the peak of migration along the Oregon Coast, for a few weeks between late December and mid-January, thirty whales per hour can pass by. By January, most whales have arrived along the Baja coast, concentrated mainly in three areas: Laguna Ojo de Liebre, also known as Scammon's Lagoon; Laguna San Ignacio; and Magdalena Bay. They will spend a couple of months there—mothers with new calves a bit more—before migrating north again from February to June. Oregon Coast whale watchers can see up to six whales per hour heading north at the peak of northward migration in the spring. Cows with new calves stick very close to shore in some areas, probably to protect the calves from attacks by orcas. The gray whale's coast-hugging migration pattern makes them the most easily observed of any whale, but it also puts them in a zone of danger from some human activities such as ship traffic and fishing operations.

Gray whales pass through or spend time in seven UNESCO biosphere reserves, one in Mexico, five in the United States, and one in Canada. From south to north, those are El Vizcaíno, Channel Islands, Golden Gate, Cascade Head, Olympic, Clayoquot Sound, and Glacier Bay-Admiralty Island. These whales stretch the scale of ecological connectedness and challenge us to think bigger in both space and time. They are a prescription for correcting the ecological myopia of our own species.

Gray whales were nearly wiped out by whaling in the eastern Pacific, as they had been in the Atlantic. Laguna San Ignacio was one of their last strongholds among the several wintering lagoons in Baja California that were attacked by American whalers. Charles Melville Scammon (1825–1911), a San Francisco–based whaling captain, was the first to penetrate this lagoon, somehow finding a route over the treacherous sandbars protecting its mouth in 1860, the year Abraham Lincoln was elected president. Within a few seasons it had been swept clean of whales. Scammon's name was already attached to another gray whale breeding area about eighty miles north along the coast, then called Scammon's Lagoon and now known as Laguna Ojo de Liebre—Eye of the Jackrabbit Lagoon.

Scammon was a contemporary of John Muir (1838–1914) and, like many men of that era, had a complex and somewhat conflicted view of American enterprise and its impact on American nature. Starting in 1869, he wrote seventeen articles for the San Francisco–based *Overland Monthly*, a popular travel and natural history magazine for which Muir also wrote. In 1874 Scammon wrote a book, *The Marine Mammals of the North-Western Coast of North America*, based on his close observations while hunting whales, elephant seals, and other marine mammals, and was elected to the California Academy of Sciences. Somehow the whaler was transformed into a whale scientist. Dick Russell writes, in *Eye of the Whale*, that "in eventually turning away from whaling, Scammon became an explorer, writer, and the foremost expert of his era on cetaceans of all kinds. I believe his personal metamorphosis marked the start of a shift in consciousness about the wondrous creatures who inhabit our oceans."

Scammon estimated the population of *Eschrichtius robustus* at thirty thousand in the mid-1800s. In 1930, biologists estimated that only a few dozen survived. The first International Agreement for the Regulation of Whaling prohibited the killing of gray whales in 1937, and with protection, their population began to rebound. In 1994 they became the first whale to be removed from the US Endangered Species List. The most recent stock assessment report available from NOAA Fisheries (2018) estimated that on the southbound migration in 2015 and 2016 there were about 27,000 eastern Pacific gray whales. The American Cetacean Society says that the eastern Pacific gray whale population "has made a remarkable recovery" and is now probably close to its original, pre-whaling size. Regina Guazzo and her colleagues from Scripps Institution of Oceanography, using an offshore hydrophone array deployed along the central California coast to listen for migrating gray whales, found that they vocalize during migration, providing a new method for tracking their migration and estimating population numbers. They write that "the dramatic recovery in population size over fifty years and the reoccupation of the lagoons in the winter and the coastal corridor during the migration gives hope that endangered marine mammals can survive with proper management and time."

A key piece of the story of gray whale recovery involved the creation of a whale reserve in Laguna Ojo de Liebre by the government of Mexico in 1971. A year later, Mexico decreed that Laguna San Ignacio was to be protected as a "Reserve and Refuge Area for Migratory Birds and Wildlife." Then in 1988, the El Vizcaíno Biosphere Reserve, a vast swath of desert land encompassing San Ignacio and Ojo de Liebre Lagoons, was established. This reserve is said to be the largest protected area in Latin America at almost 25,000 square kilometers. In 1993 Laguna San Ignacio and Laguna Ojo de Liebre were also listed as a UNESCO World Heritage Site, called the Whale Sanctuary of El Vizcaíno. Now not only were the whales protected as a species, but their critical breeding and calving areas also were protected at the highest international levels.

Or were they? In the early 1990s, Japan's Mitsubishi Corporation, which owned a controlling interest in a giant solar-evaporation salt works in Laguna Ojo de Liebre, put forward a proposal to expand salt production into the Laguna San Ignacio. A heated environmental debate ensued, lasting for nearly a decade. A long list of international conservation organizations lobbied the Mexican government in opposition to the Mitsubishi plan. Finally, in early 2000, Mexican president Ernesto Zedillo announced that the salt industry would not be allowed in the Laguna San Ignacio. Some people saw a parallel with John Muir's fight to protect the Hetch-Hetchy Valley in Yosemite National Park from damming—that battle had been lost, but the fight over San Ignacio Lagoon had been won.

~

A little more than a century after gray whales were nearly wiped out in San Ignacio Lagoon, word began to spread that fishermen there had started to encounter "friendly" whales, which approached their boats and seemed to want to be stroked and petted. The first of these encounters is said to have been in 1972 with a San Ignacio fisherman, Patricio Mayoral. Soon curious scientists and tourists began coming to Laguna San Ignacio to see the mother whales who approached boats to be touched and to introduce their newborn calves.

The "friendly whales" story shows clearly that the slaughter of whales at Laguna San Ignacio was so sudden and rapid that there was no time for natural selection and evolution to shape their behavior toward humans. There are many examples in which prey species, hunted over a long period of time by either human or nonhuman predators, have evolved an innate fear of those predators. Not so with these whales. Whales born only a few generations since their ancestors were slaughtered by Scammon's men at Laguna San Ignacio exhibited no intrinsic fear of people in boats, and instead seemed to be innately curious about them.

Some people immediately interpreted this "friendly" behavior as the whales trying to communicate with us. Dick Russell says the first encounters between whales and fishermen were a "decision to make contact with our species." Some people criticized that interpretation, saying that the idea of whales feeling "friendly," human-like emotions, or even curiosity, is "anthropocentric"—that is, that it projects human emotions onto nonhuman animals, with the implication being that they could not possibly feel such emotions. For me, it is practically the opposite: imagining that whales and other nonhuman animals don't have a rich emotional repertoire is what is anthropocentric. It is the equivalent of the Southern slave owners of Scammon's day imagining that slaves didn't have real emotions.

I visited Laguna San Ignacio twice—in the early spring of 2004 and again in 2005. I remember especially one warm, sunny morning, in the middle of the lagoon just off of the Kuyima camp, when a female and her calf approached the boat and lingered a long time. Everything seemed so calm, and time stretched into delight. The whales seemed to be in control of the interspecific communication and contact. It was not our choice. Another day our boatman crisscrossed back and forth for hours in the lower estuary, where we could see the breakers on the sandbars Scammon had crossed, and although we saw whales blowing all around, and lots of mothers and calves at a distance, none chose to approach us that morning.

It was a strange feeling to have a sense of being chosen by a mother whale to be introduced to her calf. And equally strange to feel that, despite our strong wish to be approached, if no females chose to approach us, we just had to wait. And then . . . in the exciting, ephemeral moment when a mother and calf did head for our boat, passing underneath, rising, and maybe gently bumping the boat's bottom and giving us a breath-catching moment, maybe lingering alongside to be rubbed and petted, the wonder at "why?" Why did this whale and her baby come up to us now? In my journal that night I wrote,

> Oh whale, can you feel whether
> My heart is proud, or pure? Whether
> I feel I deserve to touch you, or
> am simply amazed at such a generous gift?

Our lack of control of the situation somehow created a feeling that here I had encountered another self-realized being, equal to me or to any human. A fellow-being with whom we share the Earth—yet somehow of a different world. It held a powerful arcing of spiritual power, this encounter with another equally valid "presence" of being—but of nonhuman being. In such a meeting, the Laguna San Ignacio simultaneously shrinks to a self-contained world of immediate power and expands to a stage for a dialogue of existential significance. At the time I wrote,

> Oh whale, where are the secret
> worlds mapped on your broad back?
> Deep below the waves where I float,
> or among the galaxies of plankton?

~

In the 1970s, researchers began to notice and study gray whales that seemed to be spending their summers along the coasts of northern California, Oregon, Washington, and British Columbia rather than in the Arctic. Dr. Jim Darling began a photo-identification study of these summer whales off Tofino, British Columbia, in 1972; it has been continued ever since, and many of the individual whales he identified have now been known for decades. Tofino, on Clayoquot Sound, is today within the Clayoquot Sound Biosphere Region, a UNESCO Man and the Biosphere Programme biosphere reserve (like Cascade Head), established in 2000. Another photo-ID study was started by John Calambokidis in Puget Sound in 1984. Cascadia Research Collective, a nonprofit scientific research and education organization he founded, now maintains the most complete photo-ID catalogue of Pacific-Coast-feeding gray whales. Calambokidis and two colleagues published (in 2012) an analysis of the abundance and population structure of gray whales that spend the summer from northern California to Vancouver Island, based on 14,686 photographic identifications representing 1,031 unique gray whales and spanning the years 1998 to 2010.

How does this photo identification work? It's easy for us to understand the concept in our own species. People have lots of patterns that make individuals readily identifiable. Some have a genetic basis, like eye color; some fall into a gray area where both genetic and environmental factors play a role, like fingerprints; and some are purely environmental, a record of the history of that person's interaction with the environment, like scars. (A doctor who stitched up a big gash on my right hand that resulted from a bicycle crash when I was eight called the scar that resulted an "individuation mark," suggesting it was something to be proud of; at the time I wasn't so sure.) For gray whales, the shape of the dorsal hump and the number and prominence of the row of smaller bumps along the spine behind it toward the tail, called "knuckles," are variable and are probably mostly under genetic control, like human eye color. Patterns of natural skin pigmentation—splotches of gray and black—may also be inherited. But gray whales have lots of individuation marks that record their history and make each individual unique and identifiable with a good view or photo. Patches of whale barnacles, barnacle scars, and white scars from orca attacks or cuts from boat propellers are all unique. Because the dorsal hump region is always seen when a whale surfaces, the characteristics of that area are the most useful in building a catalogue of individuals. Using photos of surfacing whales, that's just what Darling, Calambokidis, and other gray whale scientists have done.

Carrie Newell has been studying the Oregon summer-resident population of gray whales around Depoe Bay, Oregon, since 1999. She did her master's degree on their diets in the area, scuba diving behind feeding whales, collecting samples of their watery defecations in sampling tubes, and then identifying what they had been eating under a microscope. Her conclusion: mostly mysids. Mysids belong to a group of small, shrimp-like crustaceans that brood their young in a marsupial-like brood pouch, so are sometimes called opossum shrimp.

Depoe Bay is just twenty miles south of Cascade Head, as the whale swims, and on a whale's scale it might as well be considered part of the Cascade Head Biosphere Reserve. Grays breakfasting on mysids off Depoe Bay in the morning might easily be enjoying an afternoon snack off Cascade Head on the same day, maybe porcelain crabs or ghost shrimp. They hang around the area, apparently, because of the productive nearshore feeding opportunities to the north and south of Depoe Bay, between Cascade Head and Cape Foulweather.

Carrie has contributed her whale photos to the Cascadia Research Collective's catalogue and has identified around a hundred whales that are unique to the Depoe Bay area. In her 2013 book, *A Guide to Summer Resident Gray Whales along the Oregon Coast,* she provides photos and descriptive nicknames, based on unique individual features, for seventy whales she has encountered repeatedly. Here's where scientists can wax poetic. "Eight Ball" is a "large, very dark female with a single white spot," who showed up at Depoe

Bay in 2004 and 2008 with a calf. "Blanco" is a male with predominantly white skin pigmentation over his whole body; he showed up at Depoe Bay in 2012, spent six months there, and courted "Comet," "Valentine," "Morisa," "Aurora," and many other female whales before heading south for Baja in mid-November. "Comet," a whale with a white streak of pigmented skin in front of the dorsal hump that Carrie says looks like a "comet's tail," is a regular summer resident at Depoe Bay. "Lucky" was a calf that arrived at Depoe Bay without his mother in 2012 with orca tooth-rake scars on his flukes and sides. Carrie speculates that his mother probably gave her life fighting off an orca attack as they crossed Monterey Bay in California, allowing Lucky to survive. "DD," for "dorsal dots," is a regular resident female, who has two prominent white barnacle scars near her dorsal hump. "Zebra Stripe" is a longtime summer resident with a row of white scars from an encounter with a boat propeller along his back. It's pretty clear that being able to identify and get to know individual gray whales starts to set up some personal relationships. That is certainly true for Carrie Newell. She has turned her passion and her research on gray whales into a thriving business, Whale Research EcoExcursions, based at its Whale Museum storefront along Highway 101 in downtown Depoe Bay. She and her staff take visitors offshore in inflatable Zodiacs and introduce them to the individual whales she has come to know, and nicknamed, giving them a personal encounter with a deeper, larger world.

The group of behaviorally distinct gray whales that stays along the Pacific Coast during the summer to fall feeding season has now been designated the Pacific Coast Feeding Group, or PCFG for short, by NOAA's National Marine Fisheries Service. In their 2018 population assessment for eastern Pacific gray whales, they estimated the PCFG population to number about 250 whales. The 2012 analysis by Calambokidis and colleagues found that the majority of PCFG whales seen from June to November were whales that returned frequently to the area and displayed a high degree of intra-seasonal "fidelity." Their analysis also detected some whales they called "visitors," or "apparent stragglers" from the northbound migration, which are sighted only in one year, tend to be seen for shorter time periods in that year, and are encountered in more limited areas.

Digging into the details of the 2012 photo-ID analysis presented by Calambokidis and his colleagues isn't easy, but provides some fascinating stories. Citing this study, the aforementioned NOAA stock assessment concluded that "despite movement and interchange among sub-regions of the study area, some whales are more likely to return to the same sub-region where they were observed in previous years." In other words, while some whales seemed to prefer Oregon, others liked Clayoquot Sound, others liked the San Juan Islands, and so on. Their data show the "interchange" of summer resident whales between Clayoquot Sound, where Jim Darling studied them, and the Cascade Head area, where Carrie Newell's photos document their presence. In 1998, for example, there were eleven unique whales seen only on the Oregon Coast and sixteen unique to Clayoquot, but six whales were spotted in both places during that summer. In the thirteen-year period from 1998 to 2010, 206 whales were spotted only in Clayoquot Sound, and 104 whales only in Oregon, but 51 individuals were seen in both places. "Despite extensive interchange among subregions in our study area, whales do not move randomly among areas," Calambokidis and colleagues wrote.

An as-yet-unanswered question is this: Is the behavior of the so-called Pacific Coast Feeding Group something new? It could be a new phenomenon that began in the early 1970s, caused by changing climate or ocean conditions or some other unknown factors. Or, it could be that there had been some summer resident whales all along, but no scientists had looked carefully and realized that they weren't just passing by briefly on their way to the Arctic. You don't necessarily notice or see what you don't look for. Still another intriguing possibility is that Pacific Coast summer resident gray whales have been reinventing, reexploring, and restoring behavioral patterns and diversity that may have been common before they were hunted almost to extinction in the mid-1800s. It seems perfectly plausible—no, more than that, even "eco-*logical*"—that some gray whales would have had their reasons to be laggards, to dally and delay, and *not* swim five thousand or six thousand miles to the Arctic, when they could find enough food by swimming half that far. That explanation—of a restoration of diverse, adaptive behavioral patterns—would be similar to what happened with juvenile salmon when the salt marshes in the Salmon River estuary were restored through the actions mandated by the Cascade Head Scenic Research Area. The salmon explored the restored habitat and expanded their feeding and migration opportunities, and their populations.

Researchers are trying to answer the question of whether the Pacific Coast Feeding Group is something new. At a 2017 workshop on the status of eastern Pacific gray whales, it was reported that, based on photo identifications of migrating whales seen by whale-watch naturalists in southern California, PCFG whales often migrated in groups together, in both the southbound and northbound migrations, although such groups were more common going south. Both males and females were present in these groups, and the workshop report concluded that "the association of PCFG whales on both northbound and southbound migration suggests that these animals may also associate on the winter breeding grounds. This may increase the chances that whales using the same feeding areas breed with one another, even when they migrate to a mixed-stock wintering area." This raises a question about whether Pacific Coast Feeding Group whales are not only behaviorally different, but maybe are also genetically distinct.

The genetic story is somewhat murky at the moment. One recent study found significant genetic differences in mitochondrial DNA between Pacific Coast Feeding Group whales and those in the much larger Arctic feeding population. Mitochondrial genes, contained within cellular structures called mitochondria, don't assort and combine the DNA from the mother and the father the way genes from the cell nucleus do, but are passed directly from mother whales to their calves. Because mothers lead their calves to the feeding grounds—most to the Arctic but some to their favorite areas along the Pacific Coast—it makes sense that those populations could become genetically differentiated in the mitochondrial genes passed directly from mothers to their offspring. But no significant differences were found in nuclear DNA markers—which do assort and combine genes from both parents—between PCFG whales and other eastern North Pacific whales. That strongly suggests that, during the fall migration and in the breeding lagoons of Baja, PCFG whales mix it up and mate with the Arctic-feeding whales.

The interest in this question of genetic differences between Pacific Coast Feeding Group grays, of which there are only a few hundred, and the much larger Arctic-feeding population, arises because of the issue of "evolutionarily significant units" within species. This issue of ESUs has come up earlier in discussions of salmon and silverspot butterflies. In the interest of conserving the full range of genetic diversity in a species—and thereby its evolutionary potential, adaptability, and resilience—the Endangered Species Act recognizes distinct ESUs for conservation and management purposes. If the PCFG gray whales were clearly an ESU, specific management plans for its conservation could be required, compared with the rest of the stock of eastern Pacific gray whales. The NOAA Fisheries 2018 stock assessment report for gray whales concluded that "there remains a substantial level of uncertainty in the strength of the lines of evidence supporting demographic independence of the PCFG." Are the whales that spend the summer feeding off Cascade Head or Clayoquot Sound genetically different enough from their kin who feed in the Bering Sea to be an ESU, and thus deserving of special protection? Or do they just have different family traditions, a different migratory and feeding culture? And if the latter, does that cultural tradition in PCFG grays deserve special recognition and protection in and of itself? In Canada, the government agency responsible for answering these questions, Fisheries and Oceans Canada, is closer to a decision than is NOAA Fisheries. In a 2019 report prepared under the Species at Risk Act, the Canadian equivalent of our Endangered Species Act, they designated the PCFG an ESU and its proposed status as "endangered."

~

The Makah Reservation occupies the northwestern tip of the Olympic Peninsula and has an area of only forty-seven square miles. On the north, Neah Bay looks out across the Strait of Juan de Fuca, and on the west, Makah Bay (formerly spelled Mukkaw) faces the open Pacific; the headland of Cape Flattery, with its lighthouse, and Tatoosh Island divide the two bays. About 1,400 Makah now live on the reservation, most in the only town, Neah Bay. I spent the better part of two summers camping on the shore of Makah Bay while doing fieldwork for my PhD dissertation on the black turban snail. A rocky reef protruding into the middle of the bay had a tide pool that was perfect for the research I was conducting on habitat choice. As I described earlier, it had been one of the sites for the pathbreaking work of Robert Paine on the *Pisaster ochraceus*, the ochre seastar, which led to his concept of "keystone species."

After getting up at 5:00, 5:30, or 6:00 a.m. to mark and measure snails at the morning low tide, I would go back to camp for a big breakfast, then usually launch off on an adventure somewhere within the nearby Olympic

National Park. Although I didn't pay much attention to it at the time, Olympic had already been designated a UNESCO biosphere reserve in 1976, like Cascade Head, among the first group of US biosphere reserves. One of my favorite places to go was Lake Ozette, where a boardwalk trail headed through a western redcedar swamp to Cape Alava. Just north of where the trail comes to the coast is an archeological site where, around three hundred years ago, a Makah village was buried by a mudslide, preserving even delicate basketry and fabrics in anaerobic mud. (Many of the artifacts from the site are on display at the wonderful Makah Cultural and Research Center in Neah Bay.) About a mile south of Cape Alava is an area called Wedding Rock, where more than forty petroglyphs cover rocks and boulders at the tideline. Of all of the rock art, my favorite shows two whales, with open mouths that make them appear to be smiling or talking, and two cartoon-like round faces with open oval mouths that look like they are singing. Two whales, two faces—I took these as side and front views of the same individuals, and started calling this petroglyph The Whale People. It called up such an immediate empathy that I took it as a visual statement that the ancient artist thought of whales as people, or vice versa.

The Makah are closely related, culturally and linguistically, to the Nuu-chah-nulth tribes living along the west coast of Vancouver Island, British Columbia, including around Clayoquot Sound. As it did for their neighbors to the north, whales and whaling played a central role in Makah culture and spirituality. Archeological evidence shows that these peoples have been hunting whales for more than four thousand years, and genetic identification of the whale species that were hunted shows that gray whales were the most common prey, making up nearly half of the whale bones in whaling middens. Humpback whales were next most common. "Whales provided ancient Makah people with food, raw materials, a source of spiritual and ceremonial strength," according to Anne Renker, an ethnographer of the Makah. Whaling was woven into Makah economic, political, and spiritual life. A whale hunt required physical and technical preparation, but also rituals of ceremonial bathing, abstinence, and prayer, performed both before and after the hunt.

The Makah suspended their whaling in the 1920s, mainly because there were no more whales to hunt. After gray whale populations had rebounded, and the species was taken off the Endangered Species List in 1994, the

Makah petitioned the National Marine Fisheries Service and International Whaling Commission to allow them to resume hunting whales. In May 1999, for the first time in more than seventy years, they successfully harpooned a gray whale from a cedar canoe using mostly traditional methods—although they provided the coup de grace, a shot to the whale's brain, with a high-powered rifle. Since that hunt twenty years ago, however, the US government has used the Marine Mammal Protection Act to block the Makah from hunting whales, even though the 1855 Treaty of Neah Bay gave the tribe that right. In 2005, the Makah requested a waiver to allow them to hunt a whale again. In response, NOAA Fisheries initiated an analysis under the National Environmental Policy Act, and began to prepare a draft environmental impact statement, released for public comment in 2008. But in 2012, NOAA withdrew that draft EIS and began preparing a new one, "in light of substantial new scientific information central to our consideration of the Tribe's request and our NEPA evaluation." The decision is still in legal and bureaucratic limbo. NOAA Fisheries stated, in March 2019, "It is too soon to say if or when the Tribe might resume such hunts. . . . There are many additional steps and processes in our review of the Tribe's request."

Why is it so complicated and controversial to decide whether resuming the Makah hunt would jeopardize gray whales on the West Coast? Resuming the Makah hunt would lead to a take of only two or three whales a year out of the estimated 28,000 gray whales in the eastern Pacific, and NOAA's most recent stock assessment of gray whales "estimated that over 600 individuals could be removed annually from the population without affecting the stock's optimum sustainable population size." But on the other hand, if Jim Darling's view—and that of Fisheries and Oceans Canada—that the approximately two hundred whales in the PCFG are an evolutionarily significant unit—for genetic or even behavioral reasons—then the Makah hunt could have a much larger impact.

~

In May 2019, another gray whale washed up on a beach in Point Reyes National Seashore, part of the Golden Gate Biosphere Reserve. California newspapers ran headlines such as, "As California Gray Whale Deaths Mount, Scientists Raise Alarm about Ocean Health," from the Santa Rosa *Press*

Democrat, and "Gray Whales Are Starving to Death in the Pacific, and Scientists Want to Know Why," from the *Los Angeles Times.* At least sixty-four gray whales washed up dead along the California, Oregon, and Washington coasts in 2019, as the spring migration to the feeding grounds got into full swing—the highest number since 2000. Whale researchers like John Calambokidis have been examining many of the dead whales, and have found that they had insufficient fat reserves to make it through the long winter season when they are mostly fasting, not feeding.

What's going on? Should we be worried? Is the melting of sea ice in the gray whales' Arctic feeding areas reducing the amount of food they find? Are they a canary in the coal mine of climate change? Or is this just another of nature's cycles, something we should expect in a dynamic but peaceable kingdom?

According to the gray whale stock assessment conducted by NOAA Fisheries in 2018, "Eastern North Pacific gray whales experienced an unusual mortality event (UME) in 1999 and 2000, when large numbers of emaciated animals stranded along the west coast of North America. Oceanographic factors that limited food availability for gray whales were identified as likely causes of the UME." Most of the dead whales then were adults. Calf production went back up, the gray whale population rose back to levels seen before that period of so-called unusual mortality, and it appears to have been stable since 2003.

Ecologists studying populations describe the point at which the numbers of a given species are in balance with the food or other resources available in their environment as the "carrying capacity" for the species. When the population is more or less stable, not growing or shrinking dramatically, that is evidence that it is at or near carrying capacity. Gray whales seem to be there now; their population is around that estimated by Scammon in the mid-1850s, before whaling began. When a population is close to carrying capacity, it is more susceptible to environmental fluctuations, and so its numbers will fluctuate. The fact that there was an unusual mortality event in 1999–2000 and something similar seems to be happening now should probably be taken as to be expected, and normal—not unusual—and a good sign that these whales are back up to the highest numbers the Northeast Pacific can support.

A hopeful view is that what we are seeing is the resilience of gray whales. The Arctic climate is changing significantly, resulting in a reduction in sea ice cover, and the summer range of these whales has expanded significantly in the past fifteen years. A study in the Arctic found that in response to these changing conditions, gray whale prey found in open water above the bottom is likely to increase, while amphipods and other bottom-dwelling prey are likely to decrease. Because gray whales are relatively versatile and can feed both on the bottom and in more open water, they may do fine under a changing climate. In fact, it's not clear whether a melting Arctic Ocean will hurt or help them. The oldest gray whale fossils are about 2.5 million years old, and since then Earth has experienced more than forty major cycles of warming and cooling. A 2011 study by Nicholas Pyenson and David Lindberg concluded that gray whales have been able to adapt to climate changes over millions of years by changing their migratory and feeding habits.

Behavior may enhance the adaptability and resilience of gray whales. Behavioral preferences involving migration to particular areas are transmitted by teaching and learning, a kind of cultural, rather than genetic, transmission. It is not hard to imagine that if mothers lead their calves to feeding grounds, they also teach them how and what to eat. Along the Oregon Coast, they must show their calves how to suck mysids out of kelp forests, and in the Bering Sea how to gulp amphipods off the bottom. If oystercatchers can teach their chicks how to hammer or stab mussels, certainly gray whales can transmit feeding traditions in their families.

Gray whales are a conservation success story. The five "re" words sum it up. Resistance to hunting whales and resistance to the destruction of their wintering lagoons played a role. Research has clarified their ecology and migration, although there is still much to learn about their genetic structure and social behavior. Restoration of the population to near the natural carrying capacity has occurred, and restoration of the behavioral diversity of feeding traditions may be occurring in the Pacific Coast Feeding Group. Whale watching is a growing kind of ecological tourism, introducing more and more people to the magic of the whale-world, whether in Laguna San Ignacio or Depoe Bay. You could say that reconciliation between people and gray whales is occurring. Finally, these whales are likely to be resilient, even in a time of Arctic thawing and ocean change. Their diverse diet and migratory and feeding diversity give some confidence about their resilience.

~

Homero Aridjis, one of Mexico's greatest poets and environmentalists, wrote his poem, "El ojo de la ballena"—in English, "The Eye of the Whale"—after a trip to the Laguna San Ignacio in 1999. Aridjis had founded (in 1985) and led the Grupo de los Cien—the Group of 100—an association of prominent artists and intellectuals that included Octavio Paz, Juan Rulfo, Rufino Tamayo, Gabriel Garcia Marquez, and others, devoted to environmental protection and the defense of biodiversity in Mexico and Latin America. The Group of 100 was responsible for an official decree by the Mexican government, in 1986, that ensured protection for the forests in Michoacán where migratory monarch butterflies overwinter. That area is now a UNESCO biosphere reserve, the Reserva de la Biosfera Mariposa Monarca. Aridjis also led the defense of San Ignacio Lagoon against the expansion of the Mitsubishi salt works described earlier.

His poem "The Eye of the Whale" starts with a reference to the Bible, Genesis 1:21, and begins (*en español,* then with English translation):

Y Dios creó las grandes ballenas
allá en Laguna San Ignacio,
y cada criatura que se mueve
en los muslos sombreados del agua.

And God created the great whales
there in Laguna San Ignacio,
and each creature that moves
in the shadowy thighs of the water.

The poem moves on, Aridjis imagining that "God" enjoyed seeing the whales making love and playing with their calves in what he called the "magical lagoon." He imagines the whales peeking above the wáter and seeing God among the dancing waves—God as seen through the eye of a whale, that is.

And, in the final stanza, he gives away his perspective:

Y las ballenas llenaron
los mares de la tierra.
Y fue la tarde y la mañana
del quinto día.

And the whales filled
the oceans of the earth.
And it was the afternoon and the morning
of the fifth day.

Aridjis deliberately ends the poem on the fifth day. So very telling because, according to Genesis 1:27–28, on the sixth day of creation, "God created man in his own image, in the image of God created he him; male and female created he them. And God blessed them, and God said unto them, be fruitful, and multiply, and replenish the earth, and subdue it: and have dominion over the fish of the sea, and over the fowl of the air, and over every living thing that moveth upon the earth."

And thus "man," created in the image of the god of Genesis, slaughtered the great whales his maker had supposedly created, and rendered their oil for lamps, almost causing their extinction in his supposed right and duty to subdue the Earth and have "dominion" over all its creatures. By ending his poem on the fifth day, Aridjis gives away his true opinion of man, and god, for that matter. Causing the near-extinction of the great whales was the hubris of our species, and not the wish of the god the whales saw dancing among the waves, Aridjis says to us.

In the Laguna San Ignacio, the whales are breaching and splashing in the constellations of their ocean galaxies. They are diving and playing in the deep waters of my dreams. They are calling me back. Through their eyes they see god dancing among the silver waves, and I am drawn to join them there in that world of the fifth day, a world that precedes my own species. Maybe there in that ocean, in the ocean of time, I can finally understand how we can reconcile the sixth day with all the other days, and transcend it to return to an Earth with a harmony of species, *none* with dominion.

13 *Salmon in the Net of Indra*

> *And when other men reply to the man born blind, there are diversities of color and spectators of these diverse colors; there is a sun and a moon, and constellations and stars, and spectators who see the stars, the man born blind believes them not, and wishes to have no relations with them.*
>
> —Fragment from the Lotus Sutra, published in *The Dial*, January 1844

It has started raining here at Cascade Head, finally, after an October that made the locals nervous because it was so dry, frustrated the fishermen because the salmon hadn't been called in by fresh water flushed from their natal streams, and repressed the mushrooms waiting to burst from underground mycelia.

On my afternoon run last week I passed, as usual, through the parking lot at Knight Park, where a boat ramp gives access to the lower reaches of the Salmon River only about one-quarter mile from its mouth. I'd seen the white pickup with the ODFW logo on its doors there for most of the month of October—that's the Oregon Department of Fish and Wildlife. These are the official "creel checkers" who are responsible for checking every boat coming in for tagged Chinook. The data on landings here would be combined with that from other locations in the United States and Canada as part of the monitoring of "salmon stocks of mutual concern" under the Pacific Salmon

Treaty of 1985 in order to "prevent over-fishing and provide for optimum production, and to ensure that both countries receive benefits equal to the production of salmon originating in their waters."

I'd been meaning to talk to these guys all month, and when I walked up to the driver's side of the truck, the young ODFW fisheries biologist rolled down his window, stopped texting, and greeted me. Yeah, it has been a really low year for fish, he said. Normally he'd be seeing twenty or more fish a day this time of year, the fall Chinook run, but he'd only counted seven or eight fish in the whole last week. In a good year, this parking lot would be crowded with pickups and boat trailers, but on my runs I'd seen only one or two on a given afternoon. "What's going on?" was my next question. He listed a number of possibilities. One was that the summer of 2015 had been very dry, and water temperatures in the Salmon and some of its tributary creeks got really high, almost to lethal temperatures for juvenile salmon. So, the numbers of smolt going to the ocean that year might have been low; and because most Chinook spend three years in the ocean before returning, we could now, in 2018, just be seeing the low return from the smaller-than-average 2015 population, he said. Or, maybe there are a whole bunch waiting out there in the ocean until it starts to rain; when they can smell the fresh water coming out, they'll know it's time to come home. Or maybe it's ocean conditions.

"Ocean conditions?" I asked. Yeah, he said, something going on out there. There was this "blob" of unusually warm water off the coast for a while, and maybe that held them way offshore so they couldn't come in. There have also been patches of water with low oxygen levels out there too from time to time, and maybe those kept the fish away. Or maybe it was the feeding conditions . . . or maybe . . . I got the idea there were plenty of "maybes," and not that many firm answers.

In trying to understand, manage, and harvest salmon, European and Euro-American cultures over the past few centuries have focused only on the salmon. We've treated them like just another "fish of the sea," over which we were commanded to have "dominion," according to the dominant, dominating Judeo-Christian worldview of Western culture. But salmon are more than salmon. They are part of an ecological web that reaches deep into watershed forests and far out into the ocean, linking them in a network of interconnection and interdependence.

When I was visiting the Clayoquot Sound Biosphere Region on the west coast of Vancouver Island in British Colombia, where there is a very strong Native American presence, I was told that they try to make all their decisions according to the Nuu-chah-nulth First Nations' principle of *hisuknis cawaak*, which they translated as "everything is one and interconnected." A fisheries biologist there said he thinks we need to take an "ecosystem-based approach" to salmon management, but one that starts from the salmon's perspective, not the human perspective. I wasn't quite sure what he meant, but it sounded profound, and worth pondering.

So, maybe it's not just overfishing, or hatcheries, or climate variability, or "ocean conditions," or any other one thing that is causing problems for salmon—or that caused the threats to spotted owls, Oregon silverspots, beavers, and gray whales. Maybe our Western worldview is the root cause of the problem. I said at the beginning of these essays that one of the three main lessons from Cascade Head is that worldviews matter. They are the framework or metaphor that conditions how we think about the human-nature relationship. We project that view onto the landscape (or seascape), and it shapes our individual and collective actions. But we are like "the man born blind" referred to in the Lotus Sutra, blinded to the diversity of ways we *could* view our place in nature by our culture's dominant Western worldview—by which I mean the complex of the Middle Eastern monotheisms, Greece and Rome, Europe, industrial capitalism, reductionist science, and anthropocentrism. Time to take a little philosophical field trip eastward, stopping by Walden Pond on our way to Asia, to see if we can see the "diversities of color" and "constellations and stars" that the blind man refuses to look at, searching for a more ecologically sympathetic worldview. Ready for the trip?

~

The pages of my cheap old paperback copy of *Walden* are now faded to yellow and getting a little crisp. Inside the cover page I wrote my name and the room number of my freshman dorm room, I guess so someone could return it to me if I lost it. I didn't, and now I treasure it as a source for understanding my debt to Thoreau. The marginal notes I made then are like blazes by the trails I walked with him in the Walden woods. A tiny penciled checkmark with an exclamation point stands in the margin next to, for example, "Not till we are lost, in other words not till we have lost the world, do we begin to find ourselves, and realize where we are and the infinite extent of our relations."

In the introduction to that paperback copy of *Walden,* Walter Harding, a noted Thoreau scholar and biographer, wrote that "*Walden* reaches its highest levels as a spiritual autobiography. . . . To Thoreau and to the perceptive reader *Walden* is as much a religious document as any collection of sermons." Harding describes the organization of chapters in *Walden* as "a careful alternation of the spiritual and the mundane . . . the practical and the philosophical." When Thoreau went to Walden Pond, "His purpose was profoundly religious," says his recent biographer Laura Dassow Walls.

Tracing the etymology of the word "religious" leads back to the Latin verb *ligare*: to link, bind, connect, tie. Religion is about "rebinding" or "binding back again," but the question is, to "bind back" to what? For Thoreau, it was to the depths of Walden Pond, to nature, to bedrock. In *Walden,* he counselled,

> Let us settle ourselves, and work and wedge our feet downward through the mud and slush of opinion, and prejudice, and tradition, and delusion, and appearance, that alluvion which covers the globe, through Paris and London, through New York and Boston and Concord, through Church and State, through poetry and philosophy and religion, till we come to a hard bottom and rocks in place, which we can call reality, and say, This is, and no mistake.

The second winter at Walden, Thoreau witnessed a new scene on the frozen pond whose depth he had plumbed the winter before. His reaction shows us he had a very sophisticated imagination of interconnectedness, what he had called "the infinite extent of our relations." Because the railroad passed so close to the western shore of the pond, a Boston ice company took advantage of the easy transportation to harvest Walden's winter ice, shipping it to a global market. In *Walden,* Thoreau—who had just read a translation of the sacred Hindu text, the *Bhagavad Gita*—reflected on the ice-making:

> Thus it appears that the sweltering inhabitants of Charleston and New Orleans, of Madras and Bombay and Calcutta, drink at my

> well. In the morning I bathe my intellect in the stupendous and cosmogonal philosophy of the Bhagvat-Geeta, since whose composition years of the gods have elapsed, and in comparison with which our modern world and its literature seem puny and trivial. . . . I lay down the book and go to my well for water, and lo! there I meet the . . . priest of Brahma and Vishnu and Indra . . . come to draw water . . . and our buckets as it were grate together in the same well. The pure Walden water is mingled with the sacred water of the Ganges.

It is not improbable that British colonial authorities in India were sipping gin-and-tonics chilled with Walden's ice, so Thoreau's image of the teleconnections of global trade may not have been so far-fetched.

The passage indicates the extent of global trade not just in material goods but in philosophical and religious ideas that was taking place in Concord at the time, with Thoreau at its center. When Ralph Waldo Emerson took over as editor of the Transcendentalist literary journal *The Dial* in 1842, he started a regular section called "Ethnical Scriptures" and drafted Thoreau as its editor. They began to publish translations from some of the foundational works of Hinduism and Buddhism that were just then arriving in the United States. Some were "discovered," and sometimes translated, by British civil servants working in India or Nepal—perhaps some of the same ones sipping their G&Ts on Walden ice. "In 1843 the first copy of the *Bhagavad Gita* arrived in Concord," and Emerson at first mistook it for a work of Buddhism, according to Rick Fields, writing in *How the Swans Came to the Lake: A Narrative History of Buddhism in America*. Such was the low level of understanding of the great religious traditions of Asia at the time—the *Bhagavad Gita* is a classic text of Hinduism. During his first winter at Walden, Thoreau read it. In the global trade of the time, ice from Walden Pond was traveling to Asia, and ancient ideas from Asia were beginning to arrive on American shores.

The Concord Transcendentalists, searching for ways to bridge the gap between traditional philosophy and religion and emerging science, were open to many influences. "Transcendentalism, pantheism, and related ideas were all in the air, asserting the unity of spirit and matter and each claiming that it offered the best marriage of science and the imagination," wrote Donald Worster in his 2008 biography of John Muir. Emerson and Thoreau were certainly somewhat familiar with the Greek Stoics, whose naturalistic philosophy and ethics must have interested them. They were also familiar with Emanuel Swedenborg, the controversial Swedish scientist-turned-mystic, who had studied crystal formation and argued that the seemingly automatic, self-organizing patterns generated in crystallization reflected a deeper connection and pattern in nature. Swedenborg used an analogy from the Stoics, writing in *The Principia* in 1734, that nature was like a spiderweb: "For it consists, as it were of infinite radii proceeding from a center, and of infinite circles or polygons, such that nothing can happen in one which is not instantly known at the centre, and thus spreads through much of the web. Thus through contiguity and connection does nature play her part."

The Stoic-Swedenborgian spiderweb is very similar to the Buddhist image of the Net of Indra described in the Avatamsaka Sutra. Indra's net is pictured as stretching infinitely in all directions, and at each of the knots of the net is a glittering jewel. All the other jewels in the net are reflected in each individual jewel, and each jewel reflected is also reflecting all the other jewels. This metaphor describes what was called, in Pali (the original language of the Buddhist canon), *paticca samupadda*, "dependent co-arising." Modern Buddhist teachers have called it "interbeing," or "the harmony of universal symbiosis." This is a theory of mutual intercausality, interconnectedness, and interdependence. It is a worldview from the same ecophilosophical galaxy as Alexander von Humboldt's "kosmos," the Nuu-chah-nulth First Nation's principle of *hisuknis cawaak*, and the "everything is connected" view at the heart of ecology. When Thoreau wrote that humans need to "realize where we are and the infinite extent of our relations," he had this kind of idea in mind.

We think in metaphors, often—and even scientists do. Metaphors are the templates of pattern, and having those templates helps scientists—and everyone—"see" the patterns and relationships underlying the superficial "data" of experience, which often appear chaotic. Thoreau wrote in his journal on August 5, 1851, "The question is not what you look at, but what you see." Seeing deep patterns needs a metaphoric, poetic mind.

Thoreau and his Concord colleagues were not familiar with the Avatamsaka Sutra, but as the editor of the "Ethnical Scriptures" section of *The Dial*, he was responsible for the first publication of a Buddhist text in the United States. It was an excerpt from the Lotus Sutra, published as "The

Preaching of the Buddha," in the January 1844 issue. Thoreau was always on the lookout for spiritual and philosophical texts from other cultures, and sometime in 1843 he found an article, in French, by Eugene Burnouf, a Paris-based scholar of Pali and Sanskrit. Burnouf had translated, from Sanskrit to French, excerpts from the Saddharmapundarika Sutra, or "Sutra of the Lotus of the Good Law," a fundamental scripture of Mahayana Buddhism, which he had been sent by a British government official posted at the court of Nepal. Burnouf's excerpts were the source for the article in *The Dial*. Thoreau—who learned French and several other classical and modern languages at Harvard—was long thought to have translated Bernouf's French version to English himself, but some scholars now question that view and believe the translation was by Elizabeth Peabody, who was active in Transcendentalist circles at the time.

It is hard to know how much his exposure to Hinduism and Buddhism influenced Thoreau's thinking, and how much of what he found in the *Bhagavad Gita* and the Lotus Sutra simply resonated with his own personality and intuitions. I think it's mostly the latter. Rick Fields explores this question a bit in *How the Swans Came to the Lake*. "The Concordians were at odds with their age, and they looked to the Orientals as an example of what their own best lives might be," says Fields. Thoreau himself wrote, "There is an orientalism in the most restless pioneer, and the farthest west is but the farthest east." Fields points out that Thoreau's "nature was always that of a contemplative," and reminds us that, in 1841, Thoreau wrote in his journal, "I want to go soon and live away by the pond, where I shall hear only the wind whispering among the reeds. It will be a success if I shall have left myself behind." It's not hard to imagine those words coming from a Zen monk.

For Thoreau, moving to Walden was also an attempt to experience and dwell in the present moment, to seek a kind of temporal rather than spatial "bedrock" reality. "God himself culminates in the present moment, and will never be more divine in the lapse of all the ages," he wrote in *Walden*. In other words, heaven isn't somewhere else, it's *here and now*. This is a worldview completely familiar to Zen Buddhism. How did Thoreau get there? According to Fields, "One might say that Thoreau was pre-Buddhist in much the same way that the Chinese Taoists were. . . . That is to say, he was perhaps the first American to explore the nontheistic mode of contemplation which is the distinguishing mark of Buddhism." Thoreau, in some way, seems to have been such a psychological and philosophical experimentalist that he discovered, from a base in the worldview of his Unitarian and Transcendentalist culture in Concord, the same truths that Siddhartha Gautama discovered in 500 BCE India from his base in the worldview of his princely Hindu culture. They both touched the same universal human bedrock. Thoreau cracked all of the koan, those Zen riddles with no answers, in his meditations. Nature and his neighbors, apparently, were his teachers.

The posture taken by Siddhartha Gautama, the historical founder of Buddhism, after a night spent in meditation under the "bodhi tree," or tree of enlightenment, wrestling with all the "demons" that Mara, goddess of illusion, could conjure up and throw at him, was to reach down with his right hand and touch the ground, showing that he would not be moved—that he was touching reality, touching bedrock, touching the Earth, touching truth. Thoreau touched the Earth at Walden too. He was a prophet, trying to bind us back to bedrock by asking us to see "the infinite extent of our relations."

~

Ecology and evolution provide concrete evidence of the interdependence or "interbeing" of ecological communities expressed in Buddhist philosophy. Nutrient cycles show this clearly: animals take in oxygen from the air in order to release the energy from their food, and in the process create and release carbon dioxide; plants use carbon dioxide in the process of photosynthesis and release oxygen as a waste product, for example. There is complementarity, interdependence, between plants and animals. Food chains and food webs, metaphors for the flow of energy through ecosystems, also illustrate ecological interdependence. Predators and their prey are shaped by coevolutionary forces. Wolves and mountain lions are responsible for the fleetness and grace of deer; and deer are responsible for the ferocity and stealth of their predators. Limpet-eating black oystercatchers are responsible for the camouflage of the ribbed limpet; and limpet camouflage is responsible for the sharp vision of oystercatchers.

There is now a trend toward taking a broader view of the array of possible relationships in ecological communities, not only predator-prey interactions, but also competition among species, mutually beneficial interactions, and other kinds of symbioses. Ecologists now often talk about "ecological networks" rather than just about food webs. Evolution, over eons of time, has shaped interdependent and sometimes even cooperative relationships within ecological networks. Parasites and their hosts can coevolve relationships of mutual dependence, for example. Interactions that may have begun as harmful to the host and beneficial to the parasite seem sometimes to evolve to become mutually beneficial. Lichens, reef-building corals, and the nitrogen-fixing bacteria that live in symbiotic association with roots of certain plants—like the red alder—may all be examples of this coevolution of cooperation.

Chloroplasts are the green, chlorophyll-containing structures within plant cells that can capture light energy and store it in chemical bonds. Mitochondria are the tiny "organelles" within cells that process chemical bond energy to power cellular functions. (Mitochondria have their own DNA, replicated and reproduced separately from that contained in the nucleus of the cell in which they reside. I've mentioned the use of mitochondrial DNA to study evolutionary relatedness and distinguish ESUs in several species in these essays: silverspot butterflies, beavers, seastars, and gray whales, for example.) Mitochondria and chloroplasts are ancient and important examples of the evolution of symbiotic cooperation, the result of a process called "endosymbiosis." We now know that both are derived from ancient forms of life such as bacteria or cyanobacteria that moved inside the cells of their hosts, exchanged genes with the hosts, and made themselves at home—and invaluable. The complicated history of research on endosymbiosis and an associated process, horizontal gene transfer, is described masterfully by David Quammen in his 2018 book, *The Tangled Tree: A Radical New History of Life*. The evidence for these radical evolutionary processes emerged in the 1960s and accumulated rapidly through the 1970s and 1980s as molecular genetic techniques became advanced enough to read and interpret the DNA of all major types of life, including mitochondria and chloroplasts.

The growing recognition of the ubiquity and importance of endosymbiosis and horizontal gene transfer has complicated old notions of evolutionary relatedness among organisms, which, since Darwin's day and even earlier, had been depicted as branching trees, like the genealogical trees used to show lines of descent in human families. But, it has become clear, some "branches" of the tree of life may have touched, grown together, and fused during evolution, not just diverged. A new term, "reticulate evolution," was coined—the adjective derived from the Latin word *reticulum,* referring to a net-like pattern or structure, a network—a net.

~

I gave a workshop at the Sitka Center called Cascade Head, Oregon's Biosphere Reserve: A Laboratory for Exploring the Relationship of People and Nature. It was offered on a Saturday, and there was a good turnout of a diverse group of mostly local people. I was a bit surprised that two of the participants were acupuncture practitioners from the area. What's going on here, I wondered?

When I finally had the chance to find out, I realized it made perfect sense for practitioners of traditional Chinese medicine to be interested in "exploring the relationship of people and nature," as I'd described the workshop theme. Liz Vitale, one of those practitioners, is practicing her healing arts from Neskowin, the small town tucked just on the north side of Cascade Head. She studied for four years to earn a master's degree in Chinese medicine, and classical texts were a big part of her training. She gave me a quick course in Traditional Chinese Medicine 101.

First of all, Liz said, let's start with *qi. Qi,* sometimes spelled "ch'i" or just "chi," is something like the "life force." It's very subtle, like a "breath," or "like steam coming from a boiling pot." And then there are *jing. Jing* are lines of energy; maybe you could call them "meridians" or "channels," she said. They run through the body, but don't stop at the skin, and continue out into the environment. Body and environment are not two separate things. The full term for these channels is actually *jing mai,* and those two characters together imply something like warp and weft, together weaving the fabric of life. One part of the character for *mai* is the ideograph for water, suggesting that these channels are like the network of streams and rivers in a landscape. "It's always interesting to try to explain what I've learned back into Western terms," Liz said. I could see her straddling an East-West divide, trying, somehow, to make sense of wildly divergent worldviews, and I appreciated her taking the time to try to explain it all to me.

Acupuncture is a common part of treatment in traditional Chinese medicine, and early on in the conversation I couldn't help confessing my squeamishness about needles. Needling can "help relax painful places" where the *qi* is blocked in the meridians in the body, she said. I tried my best to imagine it. It shouldn't be painful, she said; sometimes people feel "interesting sensations," sometimes they don't feel the needles at all. Still, I was glad I was just getting information and not treatment.

The issue of diagnosis seemed especially important in her practice. She described several examples of what she does when a potential patient comes to her. In Chinese medicine, you need to evaluate "all the symptoms." If you hear someone talk about having back pain, you might do something to give temporary relief, but that won't necessarily treat "the whole system," she said. She described several case studies of her diagnosis and treatment, and what struck me was their large, environmental context. It made sense, I guess, if you picture the meridians through which the life force of *qi* flows as continuing out of the body into the environment—or vice versa, coming from the outside and passing through the body. If everything is connected, healing "inner" and "outer" are part of the same process. She described one patient for whom she prescribed, "You need more sky in your life." And another who should have more contact with water, such as wading in the ocean.

I pointed out that, to me, her methods sounded pretty "intuitive," and I think she took that as meaning not scientific, or not evidence-based. "There are some problems with trying to fit East Asian medicine into a scientific system," she said, but nevertheless argued that her diagnoses and treatments are also evidence-based, rigorous, and healing. Liz saw Western medicine, based on the narrow, reductionist, cause-and-effect logic of much of Western science, as not holistic enough to capture the reality of the human-nature relationship.

I was curious about her views on "healing" on a larger scale, in watersheds, or ecosystems, say. What are the *jing*, the meridians or channels, at that scale? And what would be the equivalent of acupuncture in a forest or a river? Being at the Sitka Center, where no artistic idea is considered too crazy, and where one of my fellow residents, an installation artist, was busy wrapping the trunks of red alders on Grass Mountain with colored cloth, we joked about making an outdoor sculptural installation of giant acupuncture needles, inserting them into a stream, a rotting log, a salt marsh, or a beach somewhere to get the *qi* flowing properly again, illustrating a version of "ecological restoration" based on the worldview underlying Chinese medicine.

We talked about feng shui, sometimes called Chinese geomancy, an ancient practice that aimed to orient the forces of the *qi* to bring about harmony among the universe, Earth, and human beings. The ideograms *feng* and *shui* translate literally as "wind" and "water." Archeological evidence seems to indicate that feng shui principles were in use more than five thousand years ago in China and influenced neighboring East Asian cultures in Korea, Taiwan, Japan, and Vietnam over the ensuing millennia, as the influence of Chinese culture spread throughout the region. An important application of this ancient art and philosophy was in architecture, including landscape architecture. It was used to orient houses and whole villages in a way that maximized environmental benefits, such as having sunlight in winter, cooling winds in hot seasons, and avoiding floods, while keeping conveniently close to sources of water for domestic use and irrigation.

Chen Bixia and Yuei Nakama presented a thought-provoking discussion of the application of feng shui principles in the modern environmental context in a 2004 article in the *Kyushu [Japan] Journal of Forest Research.* Through feng shui, they wrote, "one seeks to place oneself spatially and temporally in an appropriate relation to the flow of natural processes. The way modern ecologists deal with the relationship of man and nature has been increasingly closer to that of Feng-shui, which held the Chinese ideal that man should live in harmony with nature, and that human activities should be 'designed with nature.'" In his 2008 doctoral dissertation, Chen Bixia describes the importance of trees in the feng shui worldview:

> Greenery is considered as one important criterion for good Feng Shui. It is believed that a flourishing forest can keep the Yang Earth Vein (Qi, Living Energy) to guarantee riches and honor for the family. Plants are thought to conduct good nourishing Qi throughout a place. . . . Feng Shui tradition believes that when vegetation is flourishing, auspicious energy is flowing. On the other hand, excessive deforestation disperses Living Qi, thus harming the well being of the inhabitant. Namely, the dweller's luck will disappear with tree cutting.

At the Sitka Center, tucked under giant Sitka spruces that protect us from the winds, embedded in the green of the forest, on a ridge between two creeks, and with a view of the ocean and the sound of the surf, I could feel the auspicious, healing energy of the "Living Qi" flowing all around me, the feng shui of forest and sea.

~

How did these fish get in this forest? I can understand the enterprising, entrepreneurial Pacific salmonids—Chinook, coho, and steelhead—making their way up into these coastal watersheds. But Mennonites? Somehow they ended up here, with an outdoor camp perched on a patch of high ground in a big loop of Drift Creek, in the Coast Range foothills east of Lincoln City. The camp was established in 1964 by Mennonite congregations from Albany and Sweet Home, Oregon, and is on US Forest Service land, authorized under a special-use permit. Why, I wondered?

"What about the deep divisions caused by religion?" Gary Snyder asked in *The Practice of the Wild*. "It must be said that most religious exclusiveness is the odd specialty of the Judeo/Christian/Islamic faith, which is a recent and (overall) minority development in the world. Asian religion, and the whole world of folk religion, animism, and shamanism, appreciates or at least tolerates diversity."

It's mostly true, I think, that the Middle Eastern monotheisms—Judaism, Christianity, and Islam—have viewed humans as the most important thing in the universe, specially created by "God," "in His image," and with a god-given right to dominate nature. These monotheistic religions are anthropological outliers in the spectrum of human spiritual thought. No wonder, perhaps—they were invented by largely pastoralist tribal cultures living in some of the most ecologically harsh and impoverished ecosystems on the planet. When there wasn't much there, you had to fight for every speck of grass for your flocks and every drop of water in the desert wells, and you needed "God on your side"—*your* god, not theirs. This worldview made it an easy step for the monotheisms to become political religions, tools of conquest and empire-building, as they migrated to Europe, North Africa, and Asia.

But this wasn't a harsh desert environment, here on Drift Creek, and I was curious to learn what these Mennonites were up to. I'm quite sympathetic to the historic peace churches—the Quakers, Mennonites, and Brethren—those descendants of rebels from the European Christian tradition. I attended Quaker Meeting for a long time, in both California and Colorado, at the same time I was studying Zen, and I've always felt that people who were dedicated to nonviolence and were trying to live by their consciences were kindred spirits.

Why Drift Creek Camp? "Here nature's beauty could lead the users in an experience with God's word in a setting that would enhance the experience with the Author of both the Word and the natural setting," explains a passage on its website about its history. In 2006, the camp established a nature center, whose "vision statement" says it "will foster appreciation for the important ecological contributions of the coastal rain forest, the relationships between the living and nonliving elements, and the forest's connection to the marine environment . . . [and] will raise awareness of human impacts on the forest and will educate visitors to act as stewards of this and other environments." They call it "Creation Care." That doesn't quite sound like that old commandment from that old Middle Eastern book: "Be fruitful, and multiply, and replenish the earth, and subdue it: and have dominion over the fish of the sea, and over the fowl of the air, and over every living thing that moveth upon the earth."

If the worldview of the strange species that the British biologist Sir Alister Hardy liked to call *Homo religiosus* instead of *Homo sapiens* can evolve from that biblical one into what's happening at the Mennonite's Drift Creek Nature Center, I may become a believer in us after all.

~

Most forms of Western ethics view persons as independent egos, centers of individual choice and action. But the Buddhist "dependent co-arising" view doesn't see persons in that way. Ecology and evolutionary biology don't either. The ethics that are inspired by these differing views—the Western ego-self view versus the Buddhist eco-self view—come out to be quite different. Aldo Leopold argued articulately, from an ecological perspective, for this broader view of self and community. He believed that ethics depend on the premise that the individual is a member of a community of interdependent parts. His "land ethic" enlarged the concept of this "community" to include soils, waters, plants, and animals.

When it comes to actions and lifestyles, a world of total interdependence has both a negative and a positive side . On the negative side, anything that a person does can affect the whole system. Our ego-selfish actions have a global reach. But the positive side of total interdependence is that our actions and choices, no matter how small, can send ripples of healing through the whole system.

~

The foamlines on Westwind Beach were bigger and foamier than any I'd seen before. The winds yesterday and the day before must have generated some local upwelling, filling the water with a rich brew of coacervates, those complex, soap-like organic compounds created from the breakdown of plankton proteins and carbohydrates. One hypothesis about the origin of life involves similar compounds, somehow created in the abiotic ancient ocean, foaming on some lifeless shore a few billion years ago. Today, as the foam fronts slid to a standstill and were abandoned by the seaward-slipping water that had carried them, they bent into sinuous snakes, which then separated to form letters and words in a script of the old ocean. They hooked into giant, gentle question marks on the smooth sand.

As I bent down to look closer, I saw that my face was reflected in every bubble, small or large, rainbowed in their prismatic surfaces, and so was everything around me, in all directions. I reached down and touched the sand—cold, firm, wet, and immediate.

14 *Dancing on the Shortest Day*

> *Oh our Mother the Earth, oh our Father the Sky,*
> *. . . weave for us a garment of brightness;*
> *May the warp be the white light of morning,*
> *May the weft be the red light of evening,*
> *May the fringes be the falling rain . . .*
>
> —From "Song of the Sky Loom" (a Tewa song)

When we got to the Dance House about 7:00 p.m. it was raining steadily, and long dark on this longest night of the year. A blanket covered the entrance door, and Robert held it back, explaining that we could "turn around" if we wanted to, although it was not required. Watching other people enter, it seemed to be the common ceremonial protocol—a 360-degree spin while stepping in through the doorway shows that you are leaving the regular world behind and entering into a sacred space. Robert Kentta, my host and cultural guide, is the cultural resources coordinator of the Confederated Tribes of Siletz Indians, to whom I'd been introduced through a mutual friend.

Just inside the door, a rough-hewn Douglas-fir post almost four feet in diameter held up the central beam, a trunk two feet in diameter. Two similar beams held up the sides of the peaked roof, and a cutout section in the center, covered with a roof but open-sided, let out the smoke from a roaring bonfire. Ceiling and walls were made of cedar boards, awash in warm firelight. After pausing a moment to adjust to the dimness, we moved to our seats. A walkway ran around three sides at ground level, with a bench against the drafty walls of vertical cedar boards. Then three tiers of benches, like wide bleachers, stepped down to the level of the dance floor. The space resembled a barn set halfway into the ground. The wall opposite the entry door was made of solid cedar boards and behind it were the dressing areas for the dancers, men on the left, women on the right. The bright fire blazed in a shallow, square firepit about six feet across, edged with basalt boulders and surrounded by the dance floor of cedar planks. The flames illuminated the faces on the tiered benches, which were nearly full. Robert had spread out a blanket on one to reserve some space for us. Radiant heat from the fire reached even our upper tier, but the air itself was chilly. I had prepared for a long night, wearing five layers on top and two on my legs, so I was toasty warm.

While we waited for the dancing to begin, we talked about the history of the Dance House, and the revival of the dances here at Siletz. Starting around 1870, eradicating Indian culture was US government policy, and the government's Indian agent at Siletz ordered the burning of the six traditional dance houses. But somehow Nee Dosh, the Solstice Dance, was kept alive, held in secret in peoples' homes even though they risked arrest by Bureau of Indian Affairs agents until the 1930s, when the policy of cultural repression was revoked. Serious efforts to revive Nee Dosh began in the early 1990s, when dancers and singers from Siletz visited their relatives, the Tolowa Dee-ni', in northern California, and participated in a Nee Dosh ceremony in a traditional dance house there. Back at Siletz, they started designing and building their own dance house, and in 1996, on the Summer Solstice, Nee Dosh was performed in it—the first time a full, formal Nee Dosh had been held in more than a century. Six pieces of old dance regalia collected from the Siletz Reservation that had been in a storage facility of the National Museum of the American Indian were returned to Siletz to be "danced," including a flicker-feather headband, feather dance wands, and a dentalium-shell collar and necklace.

The dancers came out twenty or thirty minutes after we arrived, men first, then the women, circling from the dressing rooms, descending the steps to the dance floor, and arranging themselves in a U-shape around the fire. Men and women of all ages, maybe two dozen of each, alternated in the line, with children—the youngest perhaps five years old—at the ends. The Nee Dosh songs were chant-like, voices rising and falling in pitch, the rhythm pounded out with a staff on the cedar floorboards by one of the lead singers. The men were shirtless, wearing animal-skin wraps like kilts and eagle plumes in their headbands, and carrying bows and quivers made of otter skins decorated with discs and medallions of abalone shell. They took turns dancing up and down the line in front of the fire, showing off their agility and prowess as hunters, sometimes drawing arrows and putting them to their bows. The dark fur quivers sucked up the firelight, and from that blackness the mirrors of shell threw spots of light like bright moons. Especially vigorous moves sometimes drew coyote-like yips and whoops from the other dancers. Sometimes a woman, or two, sometimes even three, would step out of the line to dance up and down, straight and proud, showing off a much more subtle strength than the crouching, twisting hunters. They wore white sleeveless dresses decorated with geometric designs, and aprons and necklaces of tinkling shells. Many wore traditional basket caps, and they all carried pairs of wands tipped with clusters of feathers.

The first round of dancing went on for more than an hour, followed by a long break during which community members and dancers mingled and visited. The second round finally started, and time seemed to stretch out and open a space filled with the steady rhythm of the staff pounding the floor, jangling shells, and bouncing feathers. Every five minutes or so, the firekeeper would descend the steps below the entrance door to throw some more logs on the fire, and a cascade of orange sparks would swirl up toward the smoke hole.

This Winter Solstice Dance, Nee Dosh, is a celebration of living at the "center of the world." According to the ancestors of the Siletz, when people die, they don't go to some other happier, better place. In their worldview, there is no such place. This place provides for all our needs, this place is paradise enough. Firelight threw bouncing shadows from eagle plumes in headbands and feather wands on the walls, rows of dancing shadows behind the line of living dancers, as if all the ancestors were there too, dancing this night with us, dancing at the Center of the World.

~

I grew up between four sacred mountains. Tsikomo, the Sacred Mountain of the West to the Tewa of the northern Rio Grande Valley Pueblos, was the closest and most familiar. I saw its triangular treeless south- and east-facing side every day from my bedroom window, only about ten miles to the northwest, a barometer of the seasons: white with snow in winter, green in spring and summer, tawny in fall. Across the Rio Grande Valley in the Sangre de Cristo Range was Ku Sehn Pin, the Sacred Mountain of the East, commonly known as Lake Peak. I climbed it many times from the Santa Fe Ski Basin with friends, or alone with my dog Blackie, an old black Labrador retriever. Oku Pin, or Turtle Mountain, the Sacred Mountain of the South, was about fifty miles away, near Albuquerque. Our three television stations were broadcast from that sacred peak, which everyone called Sandia Crest. Tse Shu Pin, Shimmering Mountain, the Sacred Mountain of the North, was out of sight, but we drove past it frequently on trips north to Denver. Now known as San Antonio Mountain, its gentle dome broods over the Rio Grande gorge just north of Tres Piedras (where Aldo Leopold served as a supervisor on the Carson National Forest early in his career). Those four mountains marked the center of the Tewa World. And, as it happened, the center of my world.

The Tewa have inhabited the place between these sacred mountains since around 1300, when they arrived from centers of Puebloan culture at Mesa Verde and elsewhere in the Four Corners area to the northwest, fleeing environmental degradation, climate change, and conflict. They settled at first in villages on the mesas and canyons reaching out to the east from the Jemez Mountains toward the Rio Grande Valley, then later moved down along the river to the locations of their current pueblos. When the Spanish conquistadors and settlers came, they alternately accommodated and resisted. The Pueblo Revolt of 1680 could probably be called the First American Revolution. The Pueblo tribes lost, or at least compromised, but managed to maintain their original territories and keep their cultures and languages intact. (I still have a treasured T-shirt that I bought from a hawker at a dance that says, "Pueblo Revolt Tri-Centennial.") There are now six Tewa-speaking pueblos in the northern Rio Grande Valley.

I celebrated my early birthday parties at Bandelier National Monument, one of their early settlements. My friends and I ate hot dogs and birthday cake, then ran the trails and climbed down ladders into ancient kivas, and up ladders into cave dwellings and up many more ladders into ceremonial caves.

On hikes in the canyons around Los Alamos, we often found petroglyphs of suns and animals and hunters incised on the walls of volcanic tuff. My favorite was Kokopelli, the Humpbacked Fluteplayer, interpreted as an icon for a mythical traveling salesman and Pied Piper who linked Amerindian cultures from Mexico to New Mexico and beyond. On junior high and high school basketball teams, I played against quick, tough teams from Tesuque and Pojoaque Pueblos. Sometimes my family would go to Santa Clara Pueblo on a weekend to buy big bags of wonderful tamales from the secretary at my dad's office, which my mom would freeze and ration until the next trip. Once, my dad bought an unused old flat-spring buckboard wagon seat from her husband, because it reminded him of his days growing up on a farm in Iowa. It sat on the porch at every house my parents owned, until my dad had to move to a rest home. At my dad's workplace, piles of plutonium and enriched uranium called critical assemblies were nudged and tickled into incipient chain reactions in remote underground bunkers they called "kivas," after the ceremonial chambers of the local Pueblos. For the Los Alamos scientists, the word choice was deliberate; they felt they were manipulating the mysterious forces of the universe in their kivas too.

At some point, my mom developed an interest in the local Pueblos and made some personal connections with Pueblo people, and she started taking me to dances at San Ildefonso, Santa Clara, Ohkay Owingeh (then called San Juan), Nambe, Jemez, and other nearby Pueblo communities. Pueblo ceremonies synchronize with Christian ones, especially at the Christmas season—it being the winter solstice season, after all, and all human cultures in the northern hemisphere have tuned in to that astronomical rhythm. Winter dances in the Pueblos got linked with the imposed calendar of the Spanish Catholic colonizers, and the holidays in New Mexico are filled with dances: Deer Dances, Turtle Dances, Matachines—dances on Christmas Eve, Christmas Day, New Year's Day, and Three Kings Day, January 6, when many pueblos installed their new governors, an echo from centuries ago. My memories of Christmas in New Mexico are more strongly attached to the Deer Dance, Turtle Dance, and Matachines than to Santa and the reindeer.

The dance costumes, the regalia, were always spectacular and mysterious, and obviously a source of great pride to the dancers and their families. I went to so many dances over the years of high school and when I was home from college over the holidays that they mostly blur in my memory now, but images of a particular Deer Dance at San Ildefonso are still sharp after all these years. The day of dancing had started at dawn, when the "deer" came down from the juniper and piñon hills above the pueblo. They had danced round after round, all day, the deer-men with headdresses of antlers tipped with tufts of eagle down; crowns of yucca stalks painted yellow and turquoise; heavy necklaces of silver, turquoise, coral, and shell; bundles of evergreens on armbands tied with red yarn; bells, rattles, and deer-hoof anklets clacking with each step of dusty moccasins. The front legs of the deer were sticks held like short canes, and when the dancers moved, they moved with the quick, nervous steps of deer. The drums, beating and beating, and the songs, repeating and repeating, were hypnotic. It was the last round of the day, and now the sun was slanting through the bare winter cottonwoods, about to set over the Jemez Mountains. The dancing and drumming were expectant, building toward a climax: suddenly there was a gunshot, and chaos erupted. The dancing deer broke formation and began to run, transformed into men again, and the women and girls in the watching audience around me sprinted after them, laughing, chasing their particular "dear," who, when tagged and caught, were led peaceably and happily home to a feast of posole and fry bread.

As the final round of Nee Dosh ended after midnight, I stepped into the fresh rain-washed night in Siletz, the days getting longer again, and I was home again, at home at the Center of the World.

~

The Nechesne, or Salmon River People, occupied the Cascade Head landscape at the time of first contact with Euro-Americans—probably with Spanish explorers in the late eighteenth century, and certainly with fur trappers and traders beginning in the early nineteenth century. They were a branch of the Tillamook tribe, which inhabited a stretch of coast to the north that reached beyond the Nehalem River and Neahkahnie Mountain. Like other aboriginal peoples of the Pacific Northwest, Nechesne ecology depended on the resources found in the ocean, estuaries, rivers, and forests. "By the nineteenth century these people were fully within the general frame of culture employed by Indians from southern coastal Alaska to northern California. The basic features of their culture were the use of the salmon as a primary food, dwelling in plank slab houses, using dugout canoes, excelling in basketry-making, and emphasizing wealth as an individual goal and as a prestige factor in society," according to Oregon historian Stephen Dow Beckham. Permanent villages were built along estuaries, and temporary summer shelters of mats or grasses were constructed at fishing and berry-picking sites throughout the landscape.

At least six village sites have been identified in the Cascade Head Biosphere Reserve, none fully excavated and documented by archaeologists. Village sites near the mouth of the Salmon River have been dated to as early as 1020 before present (BP). Several dozen archaeological sites along the coasts of Oregon, Washington, and northern California have been dated to between one thousand and two thousand years ago, and ten sites indicate human occupation around 3000 BP. Rapidly rising sea levels from the melting of continental ice sheets finally stabilized by around four thousand to five thousand years ago, and coastal settlements older than that could now be offshore and underwater. But even after sea levels stopped rising, some archaeologists now hypothesize that the tectonic dynamism of the Pacific Coast has obscured a much longer record of human coastal settlement. Along the south bank of the Salmon River, in the area of the restored Y Marsh and what is called Fisherman's Point on the Westwind property, Rick Minor and Wendy Grant found buried Nechesne fire hearths and charcoal that gave radiocarbon dates of between 470 and 550 years ago. The hearths were located more than two feet below the current marsh surface, and beneath a layer of sand deposited by a tsunami. The land where the former village was situated had apparently dropped by more than half a meter—a common occurrence in Cascadia subduction zone earthquakes—and then been buried by a tsunami wave. That event predates the earthquake and tsunami of 1700 AD, whose signature I've described seeing from my kayak, suggesting that major Cascadia subduction events are not uncommon.

The Nechesne and other Tillamook groups spoke a dialect of the Salishan language family and were its southernmost speakers, linguistic cousins of Salishan-speaking tribes of coastal Washington, British Columbia, and Puget Sound, such as the Nisqually, Lummi, and Bella Coola. Linguists think that Salishan languages were spoken in North America long before the arrival of Athabascan-speaking peoples around three thousand years ago. Tillamook speakers lived in a pocket of coastal territory surrounded by tribes speaking Athabascan, Kalapuyan, Chinookan, Takelma, Alsea-Siuslaw, and other languages. Tillamook is now considered an extinct language, with no living speakers.

Franz Boas (1858–1942), a towering genius in American anthropology, spent part of the summer of 1890 living with a family on Three Rocks Road, along the north side of the Salmon River just below Cascade Head. He collected a number of stories and legends from a local native informant, which were published in 1898 in the *Journal of American Folklore*. "The following traditions were collected during the summer of 1890, when I visited the Siletz Indian Reservation for the Bureau of American Ethnology in order to gather information on the Salishaw languages of Oregon," Boas wrote.

Boas learned that Cascade Head was a Nechesne vision-quest site. One story he collected described the kidnapping of four Nechesne men by people from a mysterious land somewhere on the "other side of the ocean," who arrived in a giant canoe covered with a whale skin. In the story, Boas recorded, "The man who had escaped ran up to the house calling, 'The men from the other side of the ocean have taken my brothers!' He went to the top of Bald Mountain, at the mouth of Salmon River, where he stayed twenty days fasting. Then he dreamed of his brothers. After this he returned to the village and asked all the people to accompany him across the ocean to see what had become of his brothers." Beckham wrote that "the myth tales of the Salmon River Indians further indicate that Cascade Head, a promontory rising 1,400 feet above the sea, was used as a vigil site. Children on spirit quests at puberty or individuals seeking power or vision retired to that lonely mountain top to fast, dance, and dream."

Boas's studies of indigenous cultures, from the Inuit of Baffin Island and Kwakiutl of British Columbia to the Siletz and Nechesne in Oregon, led him to view cultures as the result of a historical process of development—a dialogue between people and their environment over time. He argued strongly against a view, common at the time, that some cultures were more "advanced" than others and that cultural evolution progressed from "primitive" to "civilized" levels—with European culture presumed to be the pinnacle. His emphasis on ethnography, folklore, art, and language in the study of culture made for a holistic anthropology that was carried forward by many brilliant students, including Margaret Mead, Ruth Benedict, Alfred L. Kroeber, and Edward Sapir. Language, Boas and his students argued, is the carrier of culture.

Anthropologist Elizabeth Derr Jacobs—married to one of Boas's students, Melville Jacobs—recorded stories from a native Nehalem Tillamook speaker and informant, Clara Pearson, in 1934, and published a collection of them as *Nehalem Tillamook Tales*. One story, "Younger Wild Women," adapted and quoted below, gives the flavor of Tillamook storytelling.

~

> South Wind was traveling up the coast from Neskowin. He camped at the Little Nestucca and caught a small salmon, which he grilled on the fire and ate. He was very hungry after a busy day, transforming things. But he didn't eat the eggs. "Well," he thought, "I am all alone. I will turn those salmon eggs into two twin girls." Soon he looked, there were two twin girls. He said, "You are my daughters." He made the girls paddle his canoe. "You children do the paddling. I am not going to paddle." But the girls couldn't seem to make the canoe go straight. South Wind got mad, and threatened to beat them if they didn't paddle straight. So the two girls jumped out of the canoe and ran. South Wind felt bad, and called out "Oh, my two children! Come back to daddy!" But they would not turn around, they wanted to travel in the hills and have some fun. They were the two sisters, the Younger Wild Women.
>
> When they were a little older, they were down around Newport, and the men there had dip baskets for catching small fish. When they would get home, before their wives got to the canoes to get the fish, the Younger Wild Women would sneak over and take all of the biggest ones. Finally, they got caught, and chased away from the village by those men's wives.
>
> Those two Younger Wild Women were angry. They thought, "We have no men who care about us. We will ask everything that is growing, 'How do I look? Do I look well with my face paint?'" They asked all kinds of trees, even Willow and Salmonberry, Pine, Salal, Fir and Cedar, Hemlock and Alder. Many of them said, "You do not

look so well. That tattoo does not look pretty." Then the Younger Wild Women would jinx them. One asked Spruce. He said, "You look quite nice, you look real pretty." Then Younger Wild Woman said "All right. Now, after a while, way back [in the future], your limbs will make the best wood. From your roots women will make baskets and become wealthy. You will be the most useful of trees way back [in the future]."

They asked Hemlock, "How does my tattooing appear?" Hemlock replied, "It is no good. You both look terrible." So they jinxed Hemlock, "You will be entirely worthless. You will be of not even the slightest value for burning. You will not stand strongly. You will fall down easily, you will be brittle through and through, in fact you will be no good at all."

Then they asked Cedar, "Do you think I look well?" Cedar answered, "You look lovely, you are beautiful women." They were very happy. They said to Cedar, "Your whole body will be useful and of worth. Your bark will make fine baskets. Doctors will use it for their headdresses. Women will make skirts of it. From your body fine canoes will be made. Expert canoe builders will get rich from making your expensive canoes. Watertight buckets will be woven from your roots. Those will be the finest water buckets." They gave that Cedar a great many good qualities.

When one of the Younger Wild Women asked Salal, "Do I look nice?" he said, "You look very pretty, you are beautiful." They told Salal, "Your berries will dry well and will keep all winter. They will be excellent winter food. That is how you will be very useful."

That's why some things are the way they are.

~

Before most American Indian people on the Oregon Coast had even seen Euro-Americans, they were touched by the Old World diseases. Without any genetic immunity to diseases like smallpox and measles from a history of evolution with those diseases, and no acquired immunity from exposure during their lifetimes, they were highly vulnerable. Charles Wilkinson, in his 2010 book *The People Are Dancing Again: The History of the Siletz Tribe of Western*

Oregon, writes that, starting in the 1770s, epidemics hit Oregon tribes at least every decade until the 1850s and beyond. Many tribes are thought to have lost 90 percent or more of their former populations. For the tribes in the Siletz confederation, their population dropped from an estimated 52,000 to 4,200, a decline of 92 percent, according to Wilkinson. With no respect for age, wealth, power, or status in nonimmune populations, introduced diseases were a demographic calamity, devastating to native cultures and economic systems.

The history of the Native American tribes of western Oregon after direct contact with American traders and settlers is brutal and sad. During the early part of the nineteenth century there were not many Euro-American settlers, and there was some degree of respect between the two cultures, based partly on reciprocal economic benefits from the fur trade. But after about 1850, with an increasing tide of settlers coming across the Oregon Trail, and with the discovery of gold on the Rogue River, things turned horrible and deplorable. During the 1850s, as the US government was negotiating treaties with Indian tribes, "some federal officials honestly kept the best interests of the tribes in mind," according to Charles Wilkinson. But respect for Indian rights never really took hold, and "new personalities, hard-edged and mean-spirited far beyond the trappers or the farmers, threw the lives of the Siletz people into a chaos of insensitivity and virulent racism from which it seemed they could never emerge."

Federal policy was that American Indians should be removed from their former territories and settled in remote areas away from growing American settlements. But this was complicated by the cultural diversity of Oregon Indians; there were thirty or more tribes and bands, speaking often mutually unintelligible languages from ten different language groups. The US government's solution was to amalgamate these different tribes and create just one reservation for all of them. Joel Palmer, the Oregon superintendent of Indian Affairs, got to work to implement the policy, pressuring tribal leaders to sign treaties, cede rights to their traditional lands, and move to temporary reservations where they were thrown together with unrelated tribes. By early 1855 he had gotten agreement from the Kalapuya to relinquish their rights to the entire Willamette Valley. In November 1855, President Franklin Pierce established the Coast Reservation by executive order. It supposedly would give all of the diverse Indian groups of western Oregon a permanent territory that stretched from Cape Lookout in the north to the Siltcoos River in the south, a distance of more than a hundred miles, and inland to the crest of the Coast Range.

By various routes and means, thousands of Oregon Indians were moved to the new reservation in 1856. "The removals, some by ocean-going steamships and some by land, from the homeland occupied for thousands of years were incomprehensibly confusing and brutal," wrote Wilkinson. Some people were taken by steamship from Gold Beach, up the Colombia and Willamette Rivers, and marched westward. A layover camp was established at Grand Ronde, on the upper Yamhill River, where the Indians could be watched over by soldiers at Fort Yamhill. Other groups were marched north along the coast from Port Orford to Siletz, an inland location on the Siletz River, where the headquarters of the Coast Reservation was established. Men, women, and children made this brutal march, and many died. By 1857, with thousands of Indians still at Grand Ronde—many of whom were from inland tribes, like the Kalapuya, and weren't eager to move farther away from their beloved Willamette Valley—the US government decided to establish a second, much smaller reservation there.

In this process, cultural diversity of Oregon's First Peoples was thrown into chaos. Wilkinson explained that,

> [in moving] Indian people from their homelands to the new reservations, arbitrariness often trumped order. Just as some individuals from tribes designated for the Siletz Reservation (that is, people from the Coast and Rogue River regions) ended up on the Grand Ronde Reservation, which was intended mostly for inland tribes, so too were some people sent to Siletz while the main bodies of their tribes went to the Grande Ronde Reservation.

American Indian political organization was already complicated and delicate before the relocations. Stephen Beckham wrote, "The Tillamook, like other Indian groups west of the Cascade Mountains, probably had no superior chief or overarching political authority. The extended family, village identity, and the possession of wealth constituted the basic features of identity and power in their 'tribal' life." On top of this, some white authorities deliberately did everything they could to destroy what traditional governance structures had managed to hold on. A government Indian agent at Grand Ronde proudly described, in 1879, "I have now succeeded in entirely dissolving tribal relations among these Indians, the existence of chiefs having the effect to materially retard their advancement, and it is now often difficult to ascertain to what tribe some of the younger Indians belong, so completely have they ignored their former chiefs."

Over the next twenty years after its establishment, the Coast Reservation was reduced in size by several congressional and presidential acts. In 1865, two large sections of land were removed in the central and southern part of the reservation. In 1875, the northern and southern ends were cut off. In the north, the area north of the mouth of the Salmon River (now Tillamook County) was opened to white settlement, and in the south, everything south of the Alsea River was opened. Settlers began making homestead claims along the lower slopes of Cascade Head and in the watershed of Neskowin Creek. The Dawes Act of 1887 (named after its Senate sponsor, Henry Dawes of Massachusetts, and also called the General Allotment Act) proposed to divide communally held tribal lands into individual allotments. The tribes resisted this act of Congress until 1894, but when it was implemented, many individual Indian allotments were then sold to non-Indians, further chipping away at Native American rights and sovereignty. The final blow came in 1954, with the passage of an act of Congress that "terminated" a large number of US Indian tribes—that is, ended their federal trust status. Both the Siletz and Grand Ronde were "terminated." Because termination ended tribal sovereignty over natural resources like timber and water on reservations, some Oregon opportunists were quite happy with the idea.

"Termination radically changed the lives of many Indians in Oregon, but no one expected that the dangerous policy could be reversed. Nearly twenty years after the first termination legislation, however, that is what happened," according to the *Oregon Encyclopedia*. The first "terminated" tribe in the United States to be "restored" to tribal status was the Menominee Indians of Wisconsin in 1973. Oregon senator Mark Hatfield supported restoration of Oregon tribes, and in 1977 the Confederated Tribes of Siletz Indians became the second group in the country to have tribal status restored. The restoration of the Confederated Tribes of Grand Ronde followed in 1983.

Since their restoration as tribes, the Siletz and the Grand Ronde have been working to revitalize their traditional ceremonies and languages. The revival of Nee Dosh is one beautiful example.

~

After almost three hours of talking over fish and chips and beer at the McMenamins Lighthouse Brewpub in Lincoln City, Robert Kentta and I walked out to our cars, parked close together in the parking lot. During dinner he had mentioned that the back seat of his pickup was full of dogbane, *Apocynum cannabinum,* a native plant important to the indigenous people of western Oregon because of its fibers. He couldn't resist pulling out a stalk and showing me how it was processed. First the stem was flattened, and outer "bark" worked off with a thumbnail—some people use a grooved deer antler, he said—and then the fibers were stripped out. When twisted together, they make thread or cord that can be used for fishing lines and nets, hunting nets, bowstrings, and bags. He handed me a small twisted piece, and I pulled hard but couldn't break it. These dogbane fibers, from a native plant now on the State of Oregon's noxious weed list, were strong as nylon.

How was this discovered? I wondered. And the processing techniques? This kind of traditional knowledge had to develop, culturally, from a process of observation, trial and error, experimentation—a kind of indigenous science. Such a process eventually resulted in a body of traditional knowledge and technology that was very extensive and sophisticated and that could easily be lost as a traditional culture was lost.

Research by Donald Zobel of Oregon State University documented at least sixty-eight plant and fifty-six animal species that were used by the Salmon River people, with 308 recorded uses. Zobel's ethnographic information came from Indian informants in two anthropological studies, one by Franz Boas in 1890 and a more detailed ethnography by Homer Barnett, published in 1937. Nechesne people ate salmon, mussels, clams, chitons, crabs, seabird eggs, ducks, geese, elk, deer, sea lions, seals, and whales (scavenged, not hunted), and plants such as camas, wapato, and berries of all kinds. Fire was used to manipulate and manage habitats for plants and game animals and to drive game during hunts. Their material culture of wooden objects and woven baskets made use of dozens of species, and a dozen plants were used as medicines, including salal, deer fern, licorice fern, yarrow, and oxalis.

Biologists often say that biological diversity is manifested at three levels: genetic diversity, the diversity of genes within species; species diversity, the diversity of species; and ecosystem diversity, all of the types of ecological communities, from deserts to forests, and beaches to coral reefs, that are constituted and created by species and their genetic diversity.

But they are only partly correct: there are four kinds of biodiversity, and the one biologists often leave out is cultural diversity. Culture is an "emergent property" of biological systems, the result of parent-to-offspring (or sometimes peer-to-peer or teacher-to-pupil) transmission of information. Culture can't arise in nonliving systems; it can't exist without biology. And since culture exhibits diversity, it should be treated as a fourth level, or aspect, of biological diversity. In earlier essays I have explored the genetic diversity of silverspot butterflies, ochre seastars, gray whales, ribbed limpets, and black turban snails, and argued that such diversity is important to the ecological adaptation and resilience of those species. But I have also called attention in these essays to culture in nonhuman animals—the feeding traditions of oystercatchers and migratory traditions of gray whales, for example. In those examples, the behaviors taught by parents to their offspring seem to be adaptive, and it seems plausible that "cultural evolution" could shape behavior in adaptive ways in any animal in which culture exists, just as genetic evolution shapes genetic diversity adaptively.

Humans, whose evolution took a unique track because of runaway evolutionary feedback between genetic and cultural evolution, pose some challenges in this conceptual framework. One of the hallmarks of human cultural diversity is the diversity of language, and languages can certainly reflect local biological diversity and ecological knowledge. The Salmon River people would have named and talked in their Nechesne language about the more than 120 plant and animal species they knew and used. Claude Lévi-Strauss, the French anthropologist considered one of the "fathers" of modern anthropology, alongside Boas, called the indigenous ecological knowledge reflected in native languages "the science of the concrete." Like Boas, Lévi-Strauss argued that human minds work in the same way everywhere, and no culture is superior to any other.

I thought about my conversation with Paul Engelmeyer in the Ten Mile Creek Sanctuary, where Portland Audubon is fighting to keep old-growth forest "languages" from going extinct—the "voices of MOM's children," as I called them in an earlier essay. The argument of conservation biologists—and of the Endangered Species Act—is that the genetic languages spoken in distinct population segments of any species are valuable, irreplaceable ingredients of evolutionary resilience and need to be conserved. Wouldn't the same logic apply to human languages and the traditional ecological knowledge, the "science of the concrete," they embody?

And worldviews! What about worldviews, created by the cultural evolution of human experience in the natural world, which give guidance about how to maintain harmonious relationships with nature? Language and culture are the bearers of those worldviews. We need to know and conserve that diversity of worldviews more than ever now, as we seek reconciliation with the biosphere, and a path to resilience.

Here's an example: in Coast Athapaskan language, which the Siletz are trying to revitalize, Charles Wilkinson noted,

> *Nvn-nvst-'an* is the word for "the world." Literally, *nvn-nvst-'an* means "for you it is made," but the evocative word has much context and many layers. "It is made" means that the Creator tailored each place, each village area, to suit the peoples' every need for climate, food, water, materials, medicines, and spiritual inspiration. This is the image, the place, that is so full as to mean "paradise." There is no need for a separate heaven.

~

A pair of dice made of beaver teeth were some of the sixteen objects displayed in a special exhibit at the Chachalu Museum and Cultural Center in Grand Ronde called *Rise of the Collectors*. The ivory-white sides of the two teeth, curved like shallow letter Cs and about as long as my thumb, were incised with black lines in patterns that resembled Roman numerals or tally marks or diagrams of ancient rope ladders. The front faces of the teeth—the outside of curve—were the hard enamel that makes beaver teeth into self-sharpening chisels. That part was as orange as a fall pumpkin. According to native informants, dice like these were shaken in the hand and rolled like our modern dice, and scores kept and settled according to the now-forgotten rules of forgotten games.

These dice, and the other objects displayed here, were on loan to the Confederated Tribes of Grand Ronde from the British Museum. What? Of course, that's a story in itself, well documented by historian Stephen Dow Beckham in the text of the catalogue of the exhibit, *Rise of the Collectors.* The objects were collected by Reverend Robert W. Summers, who was born in Kentucky, in 1827, and crossed to Oregon on the Oregon Trail in either 1853 or 1854. Summers returned to Missouri after his first sojourn in Oregon, married there, was ordained as an Episcopal priest, and sent to a congregation in Seattle. Then, for unknown reasons, he was removed from that congregation and posted to a church in McMinnville, Oregon. Both Reverend Summers and his wife, Lucia, a talented botanist, were fascinated by American Indian cultures, at a time when those cultures were being deliberately and cruelly suppressed by US government policies. Starting in 1871 and continuing until 1895, Summers and his wife traveled annually to Grand Ronde and other Native American communities, buying and collecting objects and artifacts, a process that Summers thought of as preserving the artistry of a vanishing people. He eventually acquired more than 130 objects for his collection. The beaver-teeth dice on display at Chachalu were one of two sets that he acquired at Grand Ronde. Summers was careful to document ethnographic information associated with the objects, and about one pair, acquired from the wife of "Whisky Jim," he wrote in his journal, "She was careful to explain when we found her beaver-teeth dice that she 'did not gamble, but other Reservation Indians did, some of them immoderately.'"

Summers moved to San Luis Obispo, California, in 1895, and died there in 1898. In a twist of fate, his collection of Oregon Indian artifacts passed, upon his death, to a British Episcopalian priest in California, who then donated it to the British Museum, where it has languished in storage until finally brought to the light again, on loan to the Chachalu Museum in Grand Ronde.

Gambling and betting games are part of many traditional Native American cultures, but the development of gambling operations to generate revenue for tribes did not start until the late 1970s and early 1980s. Several tribes in Florida and California opened bingo parlors in those years, and soon high-stakes Indian bingo operations spread to other states, competing with state-regulated non-Indian lotteries and casinos. Issues of tribal sovereignty came to the fore, of course, and soon cases were working their way to the US Supreme Court. In several cases the court decided in favor of the tribes over the states, and American Indian gaming quickly became a popular industry on reservations. But tensions and controversies increased, until finally, in 1988, Congress passed the Indian Gaming Regulatory Act, which provides the regulatory framework for tribal gambling operations. Several hundred tribal casinos, operated by more than two hundred tribes, now generate revenue totaling more than $30 billion dollars a year; these are arguably the major source of funds for self-sufficient economic development for Native Americans today.

Both the Siletz and Grand Ronde Tribes operate casinos, and visiting them had been on my "to-do" list since I arrived at the Sitka Center. But when it came to a choice between watching whales, walking beaches, or exploring old forests, I always chose "nature" over casinos. Time was finally running out, and after visiting the Chachalu Museum in Grand Ronde, I decided I'd better seize the day and stop at Spirit Mountain, the gigantic tribal casino on the edge of the tiny town. The huge parking lot was surprisingly full on a weekday in December, and the gaming tables and electronic slot machines were busy. Spirit Mountain does about one million dollars of business *per day*, I had been told. I felt like a tourist, a voyeur, a spy, and almost expected to be approached by security personnel and tossed out as I wandered around just looking and snapping a few photos.

Emboldened by my foray into the Spirit Mountain Casino, I went a few days later to the Siletz Tribe's casino, Chinook Winds, in Lincoln City. A sprawling parking lot overlooking the beach and ocean was full. I went into the dark, smoky "dance house" of the casino, and it was packed. I again felt a little foolish as I walked up and down the crowded aisles between the tall electronic gaming machines, each one pulsing and flashing with bright colors and themes. Almost every seat was taken. It was the same demographic as in the casino at Grand Ronde—generally older folks, often overweight, some looking a bit unhealthy, rough, and scruffy. They all seemed obsessed, focused only on their screens and games. For them this must be entertainment or escape—but what were they escaping? Again, I just gawked and snapped a few pictures before leaving.

I downloaded my photos when I got home, and only then saw her face. One of my quick snapshots had gotten a row of gamers, faces lit by glowing screens, most of them oblivious to me. But somehow, she, an older woman, was looking straight at me, into the camera. I hadn't stopped to look at people in real time, feeling a bit nervous and rushed. But she had seen me, photographing the scene. Her face and expression, captured in the photo, had many dimensions—maybe a bit curious, maybe a little desperate, a touch defensive, and certainly sad, with what seemed a longing for the answer to a question that the slots couldn't answer: What does all this mean?

I was immediately struck by the contrast between the expressions of Native American women in historical photos I had seen in the Chachalu Museum and this lost look my camera had captured in the Chinook Winds Casino. Those faces of the old Oregon Indian women, like Emma John from the Takelma Tribe, were care-lined and sad in their own way, but they were proud and strong; I can't find a hint there of the helpless desperation that showed so clearly on the face in my photo from Chinook Winds. Emma John and the other Indian women have a look that says "we know."

These contrasting portraits are haunting. What have we lost? What have we gained? In that same old book that instructed our species to "have dominion" over nature, I think I remember reading some wise words once, something like, "What profit it a man if he gains the whole world, but loses his soul?" Or hers, of course. Maybe somehow we have "gained the whole world" but lost our souls. How can we get them back? What do we owe to the world, to get them back again?

~

I parked in front of the Tribal Cultural Center in Siletz as a light rain was starting. Robert Kentta wasn't there yet, but he showed up soon in his red pickup. He punched the security code into the keypad at the door of the windowless collections building, and we entered the large, two-story room filled with shelves holding baskets. Lots and lots of baskets. Basketry was a prominent art and craft of the diverse tribes that were thrown together in the Coast Reservation in 1855, and this room full of baskets is an important and symbolic part of the restoration and revitalization of the culture of those tribes that is happening in Siletz now. Some of these baskets dated back to the mid-1800s, when the Coast Reservation was established, Robert said, but most were from the 1980s and later, made by some of the last of the older basketmakers.

The baskets were of all sizes, from tiny ones that could be held in one hand to baskets big enough for a person to sit in. On the sides of big storage baskets, patterns of overlay-weaving made with dark-brown fern stems suggested trees, jagged bolts of lightning, birds, or reflections of hills in water. A line of elk, also patterned in fern stem, walked regally across the bottom of a purse-like, flattened basket with a shoulder strap.

A small cylindrical basket about five inches in diameter and eight inches high was wrapped with a stylized design of a condor. "Condors in Oregon?" I asked, somewhat surprised. Yes, it turns out that the California condor was once common on the Oregon Coast, scavenging dead whales, seals, and sea lions along the beaches. The last condor was seen in Oregon in 1904. The design—a rake of wing feathers and large head protruding on a skinny neck—reminded me of the way the Andean condor was depicted in a weaving I bought in the Bolivian highlands.

A cooking basket, so tightly woven that it was waterproof, was one of my favorites. It had a striking design: an overlay-weave of sunbleached beargrass formed a pale band across the midsection of the basket, and above and below it, simple groupings of triangles reminded me of the sun setting over the ocean and its reflection on the water. Fire-heated rocks were placed in these waterproof cooking baskets with water and food—elk, salmon, mussels, wapato, camas root, berries—to cook a stew. What kind of cooks would demand this level of art in their cookware?

The basket room at Siletz gave a glimpse into the world of the culture, indigenous to this place, that wove its worldview into everyday utilitarian objects. For me, it was humbling and unsettling. I thought about my own culture, which touts its exceptionalism and advancement over all past iterations of the human experiment, and I didn't feel proud.

W. Richard West Jr., the founding director of the National Museum of the American Indian, interviewed Geneva Mattz, a Yurok basketmaker from northern California, in 2004. He wrote a delightful and thought-provoking description of how she taught her craft to a younger generation. The first step was learning basketmaking songs, but after three weeks of that, her students started to complain, she told West. "When are we going to learn to make baskets?" they asked her. She was somewhat taken aback, and explained that they *were* learning to make baskets—but first they had to learn the songs needed to ask permission of the plants they were going to harvest so that the plants would not be insulted and angry. Finally, they learned the songs and went out to harvest the grasses, stems, and roots they would use. Then, back in the classroom, Mrs. Mattz began to teach them *more* songs, this time songs about how to split, chew, soak, and otherwise prepare the materials before starting to weave the basket. The students again complained about only learning songs. West concludes the story, saying,

> Mrs. Mattz, perhaps a bit exasperated herself at this point, thereupon patiently explained the obvious to them: "You're missing the point," she said, "a basket is a song made visible."
>
> I do not know whether Mrs. Mattz's students went on to become exemplary basket makers. What I do know is that her wonderfully poetic remark—which suggests the interconnectedness of everything, the symbiosis of who we are and what we do—embodies a whole philosophy of Native life and culture.

The song is a metaphor for the basket, the basket a metaphor for the biosphere. The warp—a strong, ecocentric worldview—comes first: the hazel-stick framework of the basket. Around that we can weave the weft and overlay the pattern: the history, stories, and personal experiences of resistance, research, restoration, reconciliation, and resilience. The symbiosis of who we are and what we do.

15 *Out Beyond the Eagle's Kitchen*

> *The passion caused by the great and sublime in nature . . . is Astonishment; and astonishment is that state of the soul, in which all its motions are suspended.*
>
> —Edmund Burke, *On the Sublime,* 1756

Above the three pillars of rock where the Salmon River spills into the ocean below Cascade Head are a few old Sitka spruces where bald eagles often perch, watching over Westwind Beach, the estuary, and the Three Rocks.

"We call it the Eagle's Kitchen," Duncan said, "because that's where they hang out, waiting for whatever is on the menu for the day." And it's true. Last week, for example, I saw a big juvenile eagle patrolling the surfline on the beach, and occasionally plunging down to grab something and pick away at it. I couldn't see what it was catching, but when I walked over to the spot where it had last been, I found the remains of a Dungeness crab about four inches across the carapace, part of breakfast on that particular day.

Duncan Berry mentioned that in the course of a conversation about how he gets out to the dramatic western cliffs of Cascade Head, where the Nechesne went for vision quests, and where the conspiring conservationists took Senator Packwood on that little hike in 1973. The hike where the eagle flew up in his face as he looked over the abyss, and gave him the vision to introduce national legislation that protected Cascade Head as a Scenic

Research Area and laid the foundation for its designation as a UNESCO biosphere reserve.

Following what I thought was Duncan's description of the easiest route out to the Pinnacle, I hiked along a closed, disused shoreline road that went seaward from the River House, the private events center of the Cascade Head Ranch community. Soon the road petered out, and I was forced down onto the shore, a slippery mix of small rocks and cobbles. It was high tide, and finally the rocky shore closed out among big driftwood logs at the base of a cliff. A muddy track climbed steeply up through the alder and sword fern clinging to the slope to the north. Scrambling and slipping, pulling myself up on ferns, alder stems, and spruce roots, in a quarter of an hour I emerged on a more level bench above the sea cliff, the backs of my hands bleeding in several places where branches had stabbed off the skin without my even noticing at the time. Deep elk prints led across a wet swale under mature alders.

"Then just follow the elk trails above the Kitchen, and on out," Duncan had said, casually.

At that point I was seaward of the beach and river mouth. Giant surf from the recent storm was breaking beyond the Three Rocks and surging over them, then regrouping to break on the beach again. No eagles in the Kitchen spruces right now. They all must have been out looking for dinner. The sun, at about 3:30 p.m., was sliding under an offshore cloud layer to the southwest. I pushed on, mesmerized by the luminous light and the roaring surf. It felt wild.

Maybe this was the "easy way" to get out on the headland, but as the light faded, I decided that the return trip would be much easier and safer if I climbed up to the Nature Conservancy's trail. I pushed uphill. Cresting a ridge, I could look out toward the Pinnacle. A big herd of elk was spread out across the opposite slope, each animal casting a long shadow across the meadow in the low sun. With binoculars, I counted seventy-two. Then I hustled up the ridge, aiming for some scattered spruces. When I pushed through a gap in the trees on an elk trail, I came face-to-face with a small group of them. I froze, and they raised their heads and stared. We looked at each other from about twenty-five feet apart for half a minute, and they seemed to be trying to figure out what they were seeing. Staring at me straight-on, their expressions said "Hunh? Who are you? What are you doing here?" In the days of their ancestors, when Nechesne hunters were stalking them here on these same headlands, they wouldn't have been so clueless or slow. In the intervening generations they seem to have forgotten the former predator-prey relationship of our two species, and lapsed into a peaceable détente.

~

The word "sublime" is not used much these days, but it would have been 150 years ago. America's westward expansion and the Civil War, embedded in the context of global exploration and a rapid expansion of science, were challenging the foundations of the old worldview Americans had carried from Europe. The word sublime was then most closely associated with the New England Transcendentalist philosophers and poets, the landscape painters of the Hudson River School and their Luminist cousins, and the nature writers arguing for a spiritual value for wild nature and for its preservation.

Thomas Cole, the landscape painter, for example, described being in a wild setting and feeling "overwhelmed with an emotion of the sublime such as I have rarely felt. It was not that the jagged precipices were lofty, that the encircling woods were of the dimmest shade, or that the waters were profoundly deep: but that over all, rocks, wood and water, brooded the spirit of repose, and the silent energy of nature stirred the soul to its inmost depths."

The concept of the sublime that Cole and his contemporaries were exploring resonates with me now, and always has. Being on a jagged headland over the raging sea, or in the high mountains—exploring off-trail, with the winter light fading fast—has always been for me a blend of exhilaration and fear. Wild landscapes have always given me an immediate, felt sense that "nature" doesn't give a damn about me. The big, old world we can see in wild places doesn't care whether we live, or die, or even exist.

But somewhat paradoxically, the psychological effect of being in wild nature has always been a feeling of a much-expanded sense of self, and of the larger meaning and purpose of life beyond me. Wilderness has always "put me in my place"—a much bigger, older abode than the sometimes hassle-filled scene of my usual life. For me a wild place, with scarce traces of my own species, always gives a sense of a deep, interconnected, cosmic meaning and a feeling of well-being and calm. I can imagine—or at least I try to imagine—that for some people the feeling of being so tiny and insignificant in a world completely oblivious to our presence would be uncomfortable, even terrifying. But for me, wild nature has always felt like my true home.

Wandering out beyond the Eagle's Kitchen, I watched the sun sink sublimely over the wild sea, under the astonishing sky. This is the view they see every day.

Acknowledgments

As I look back on how this book came to be, I'm reminded of my deep debt to the many people who helped me learn all I could about Cascade Head in a very short time. I had never met most of them when I arrived at the Sitka Center for Art and Ecology in October 2018, but over the next few months they became friends and colleagues. That experience catalyzed this writing, but it grew from deeper roots, a lifelong relationship with Oregon and the Oregon Coast, for which I owe thanks to many others—family and friends, teachers and mentors.

I am grateful to the Sitka Center for selecting me as the Howard L. McKee Ecology Resident, and to the McKee family for their support of this ecology-focused residency at Sitka. Thanks to Alison Dennis, Carrie Hardison, Sara Haug, Mindy Chaffin, and Kimberly Ota—staff at Sitka—for their friendly and competent help and support during the residency. Creative collaborations and discussions with my fellow residents—poet Kim Stafford, photographer Kristen Densmore, musician Soraya Perry and installation artist Sadie Sheldon—provided enjoyable opportunities to explore the relationship of ecology and the arts.

Many people with long and deep knowledge of the place welcomed and facilitated my research and story-gathering, leading me on field trips or giving interviews, and these essays are in large part just my repackaging of the information, insights, and inspiration I gleaned from them. This is their book; I can't thank them enough. Many of those people also read drafts of what I was writing, fact-checking my descriptions and offering their comments, suggestions, and encouragement. Deepest thanks to Jerry Franklin, Paul Katen, Robert Kentta, Frank Boyden, Jane Boyden, Anne Squier, Debbie Pickering, Kami Ellingson, Stephen Dow Beckham, Nora Sherwood, Paul Engelmeyer, Conrad Gowell, Matt Taylor, Duncan Berry, Deborah Wilkins, Sarah Greene, Graham Klag, Peyton "Pete" Owston, Leah Tai, Rebecca Flitcroft, Todd Wilson, Rob Pabst, Jenna Sullivan, Dan Twitchell, Robert Michael Pyle, Kim Stafford, Pat Mangan, Liz Vitale, Kathleen O'Malley, Dan Bottom, Kathleen Dean Moore, Ray Touchstone, Fred Swanson, Julia Jones, Dick Vander Schaaf, Jim Darling, Diane Bilderback, and Michelle Dragoo. Tim Hogan, Eric Jones, Dugald Owen, Shelley Roberts, Anne McHugh, and Sue Tank also read some of these essays and made helpful suggestions. Cinamon Moffat and Keith Matteson facilitated the seminar presentation at the Hatfield Marine Science Center that gave an early test to some of these ideas. Lindsay Aylesworth, Cristen Don, Jon Putnam, Cliff McCreedy, Paul Robertson, and Barbara Triplett also deserve thanks. Thanks to Curt Meine and Peg Herring for their careful reading and constructive comments on the entire manuscript. Michael Sweeney, Jodi Lorimer, Susan Bunsick, Anya Byers, and James Ashby played special roles in supporting this work. Thanks to Kim Hogeland, Marty Brown, and Micki Reaman of the OSU Press, for shepherding the process of review and publication, and to Susan Campbell for her sharp-eyed copyediting. Duncan Berry offered his stunning aerial photograph of Cascade Head for the cover. Nora Sherwood's beautiful frontispiece, map, and chapter heading illustrations grace the book, and working together on these deepened our art-and-ecology collaboration that began at Sitka. Nora deserves special thanks, and we thank the generous donors who helped defray some of the cost of producing the illustrations.

The comparative perspective gained from visits to two Canadian biosphere regions was eye-opening and important. Thanks to Pam Shaw of Vancouver Island University for facilitating my visit to the Mount Arrowsmith Biosphere Region. Rebecca Hurwitz, executive director of the Clayoquot Biosphere Trust, arranged a rich menu for my time at the Clayoquot Sound Biosphere Region, and I also thank Laura Loucks, Faye Missar, Tammy Dorward, Tawny Lem, and Marc Labrie there. Eleanor Haine-Bennett provided a big-picture understanding of the Canadian UNESCO Man and the Biosphere Programme in a phone interview.

For their help in learning about UNESCO biosphere reserves in Ukraine, I thank my colleagues Alena Tarasova and Olga Denyshchyk, and we thank Viktor Gavrilenko, director of Askania-Nova Biosphere Reserve, and Sergii Mykolayovych Zhyla of the Polisky Nature Preserve. I'm grateful to Juanita Ames for providing a base in San Ignacio, Mexico, from which to explore the El Vizcaíno Biosphere Reserve.

The late Bill Arbus, director of Camp Arago (later called Terramar Field Science Facility), provided an outlet for my outdoor-naturalist passions at a formative time in my life. I'm grateful to my graduate professors and mentors, Jeff Mitton and Dave Armstrong, for their guidance and solidarity; to Robert Paine for encouraging me to work at his research site at Mukkaw Bay, Washington; and to Chuck Baxter for enabling my dabbling in "Doc" Ricketts Great Tidepool in Pacific Grove, California.

These essays also spring from deeper sources that deserve acknowledgment. My grandfather, Harry Sweeney, introduced me to the magic of Oregon tide pools at a young age. My father, Cleo Byers, passed on his love of science, and my mother, Mary Sweeney Byers, kept us going back to Oregon from New Mexico most summers and encouraged my curiosity in all things. My daughter, Anya, and son, Jonathan, are carrying on the traditions of science and love of nature I was taught, to my thorough delight. Beak to beak, nose to nose, that's the way it happens.

Sources

More than fifty people provided information, ideas, and insights for these essays during interviews, field trips, and informal conversations, during which contemporaneous notes were recorded in my notebooks. Newspaper articles and photos from the personal files of Anne Squier provided rich information about the events leading to the designation of Cascade Head as a US Forest Service Scenic Research Area. Letters and papers relating to the establishment of the Neskowin Crest Research Natural Area were viewed and photographed in archives at the headquarters of the Cascade Head Experimental Forest in Otis, Oregon. Online sources were all accessed most recently in January 2020 unless otherwise noted. The books, scientific articles, and websites listed below are sources I used, but this list is not meant to be exhaustive or fully comprehensive.

1. Eagle's View

Dietrich, William. *The Final Forest: The Battle for the Last Great Trees of the Pacific Northwest.* 1992. New York: Penguin Books.

Heisman, Rebecca. 2018. "Bald Eagle, The Ultimate Endangered Species Act Success Story." American Bird Conservancy. May 24. https://abcbirds.org/bald-eagle-the-ultimate-endangered-species-act-success-story/.

Snyder, Gary. 1990. *The Practice of the Wild.* San Francisco: North Point Press.

2. Big Brother in Pixieland

Overton, Sharon. 2017. *As Natural as the World Will Allow: Stories of Nature, Culture, and Transformation from the Sitka Center for Art and Ecology.* Otis, OR: Sitka Center for Art and Ecology.

PdxHistory.com. 2017. http://www.pdxhistory.com/html/pixieland.htm. Last modified April 8.

Ricketts, Edward F., and Jack Calvin. 1939. *Between Pacific Tides.* Stanford, CA: Stanford University Press.

USDA Forest Service. 1972. *Cascade Head—Salmon River Land Use and Ownership Plan.* July. Washington, DC: USDA Forest Service.

Wikipedia. "Pixieland (Oregon)." 2019. https://en.wikipedia.org/wiki/Pixieland_(Oregon). Last modified May 6.

3. The Midnight Forest and the Spruce Goose

Beckham, Stephen Dow. 1975. *Cascade Head and the Salmon River Estuary: A History of Indian and White Settlement.* Report to the US Forest Service to comply with requirements of the Antiquities and Historical Preservation Acts. Siuslaw National Forest, Hebo Ranger District.

Forest History Society. nd. "Thornton T. Munger." https://foresthistory.org/research-explore/us-forest-service-history/people/scientists/thornton-t-munger/.

Joslin, Les. nd. "Thornton Munger (1883–1975)." 1918. *Oregon Encyclopedia.* Source of Munger quotes. https://oregonencyclopedia.org/articles/munger_thornton_1883_1975_/#.XHwp18BKh0w. Last modified March 17.

Munger, Thornton T. 1944. "Out of the Ashes of Nestucca." *American Forests,* July 1944. https://andrewsforest.oregonstate.edu/sites/default/files/lter/pubs/pdf/pub828.pdf.

Steen, Harold K. 2004. *The U.S. Forest Service: A History.* Durham, NC: Forest History Society.

US Congress. 1883. Quoted from *How the U.S. Cavalry Saved Our National Parks,* https://www.nps.gov/parkhistory/online_books/hampton/chap4.htm.

USDA Forest Service. nd. "Preparing for the Future: Forest Service Research Natural Areas." https://www.nrs.fs.fed.us/rna/local-resources/downloads/rna_fs_503.pdf.

Williams, Gerald W., and Miller, Char. 2005. "At the Creation: The National Forest Commission of 1896–97." *Forest History Today* (Spring/Fall): 32–41.

Worster, Donald. 2008. *A Passion for Nature: The Life of John Muir.* Oxford: Oxford University Press. Source of quote about Sargent's views expressed in *Garden and Forest.*

4. Old Orchard and the Biosphere

Cascade Head Biosphere Reserve. 2016. *Periodic Review for Biosphere Reserve.* Final September 25. UNESCO-MAB.

Golley, Frank B. 1993. *A History of the Ecosystem Concept in Ecology: More Than the Sum of the Parts.* New Haven, CT: Yale University Press.

IUCN Red List. nd. "Przewalski's Horse." Text summary. https://www.iucnredlist.org/species/7961/97205530.

Leopold, Aldo. 1991. *River of the Mother of God and Other Essays.* Edited by Susan L. Flader and J. Baird Callicott. Madison: University of Wisconsin Press.

NASA. 2007. "Blue Marble—Image of the Earth from Apollo 17." November 30. https://www.nasa.gov/content/blue-marble-image-of-the-earth-from-apollo-17.

UNESCO. 1993. "The Biosphere Conference: 25 Years Later." https://unesdoc.unesco.org/ark:/48223/pf0000147152.

Wikipedia. 2020. "Askania-Nova." https://en.wikipedia.org/wiki/Askania-Nova#The_biosphere_reserve. Last modified January 5.

5. Reference Marsh

Bottom, Dan. 2017/2018. Restoring the River Salmon. Sitka Center for Art and Ecology.

———. 2017. Part 1. "Estuary Habitat." December 12. https://www.sitkacenter.org/journal/restoring-the-river-salmon-estuary-habitat.

———. 2018. Part 2. "Life-History Diversity and Resilience." January 9. https://www.sitkacenter.org/journal/restoring-the-river-salmon-life-history-diversity-and-resilience.

———. 2018. Part 3. "The Coho Return." February 21. https://www.sitkacenter.org/journal/restoring-the-river-salmon-the-coho-return.

Ellingson, Kami S., and Barbara J. Ellis-Sugai. 2014. "Restoring the Salmon River Estuary: Journey and Lessons Learned Along the Way 2006–2014." US Forest Service. https://www.fs.usda.gov/Internet/FSE_DOCUMENTS/fseprd563410.pdf.

Flitcroft, Rebecca L., et al. 2016. "Expect the Unexpected: Place-Based Protections Can Lead to Unforeseen Benefits." *Aquatic Conservation: Marine and Freshwater Ecosystems* 26 (Suppl. 1): 39–59. https://onlinelibrary.wiley.com/doi/abs/10.1002/aqc.2660.

Frenkel, Robert E., and Janet C. Morlan. 1991. "Can We Restore Our Salt Marshes? Lesson from the Salmon River, Oregon." *Northwest Environmental Journal* 7:119–135. https://andrewsforest.oregonstate.edu/sites/default/files/lter/pubs/pdf/pub1273.pdf.

Jones, K. K., T. J. Cornwell, D. L. Bottom, L. A. Campbell, and S. Stein. 2014. "The Contribution of Estuary-Resident Life Histories to the Return of Adult Coho Salmon Oncorhynchus kisutch." *Journal of Fish Biology* 85:52–80. https://onlinelibrary.wiley.com/doi/abs/10.1111/jfb.12380.

Montgomery, David R. 2003. *King of Fish: The Thousand-Year Run of Salmon.* Cambridge, MA: Westview Press/Perseus.

Oliver, Mary. 1990. *House of Light.* Boston: Beacon Press.

Oregon State University. 2016. "Subduction Zone Earthquakes off Oregon, Washington More Frequent Than Previous Estimates." OSU Newsroom, August 5. https://today.oregonstate.edu/archives/2016/aug/subduction-zone-earthquakes-oregon-washington-more-frequent-previous-estimates.

Schindler, Daniel E., Jonathan B. Armstrong, and Thomas E. Reed. 2015. "The Portfolio Concept in Ecology and Evolution." *Frontiers in Ecology and the Environment* 13:257–263. https://esajournals.onlinelibrary.wiley.com/doi/abs/10.1890/140275.

Schulz, Kathryn. 2015. "The Really Big One." *New Yorker*, July 20. https://www.newyorker.com/magazine/2015/07/20/the-really-big-one.

Traver, Tim. 2006. *Sippewissett, or, Life on a Salt Marsh.* White River Junction, VT: Chelsea Green.

USDA Forest Service. 1976. Final Environmental Statement for the Management Plan: Cascade Head Scenic Research Area. Corvallis, OR: USDA Forest Service, Pacific Northwest Region, Siuslaw National Forest. https://www.fs.usda.gov/detail/siuslaw/landmanagement/?cid=fsbdev7_007215.

Volk, Eric C., Daniel L. Bottom, Kim K. Jones, and Charles C. Simenstad. 2010. "Reconstructing Juvenile Chinook Salmon Life History in the Salmon River Estuary, Oregon, Using Otolith Microchemistry and Microstructure." *Transactions of the American Fisheries Society* 139: 535–549. https://afspubs.onlinelibrary.wiley.com/doi/abs/10.1577/T08-163.1.

6. Art and Ecology at the Otis Cafe

Audubon, John James. *The Birds of America*. Plate 427, White-legged Oyster-catcher. Museum of Fine Arts Boston. https://www.mfa.org/collections/object/the-birds-of-america-plate-427-white-legged-oyster-catcher-369295.

Byers, Bruce A. 1989. "Habitat-Choice Polymorphism Associated with Cryptic Shell-Color Polymorphism in the Limpet *Lottia digitalis*." *The Veliger* 32:394–402. http://www.brucebyersconsulting.com/wp-content/uploads/2012/07/Limpet-Shell-Color-and-Habitat-Choice-Polymorphism.pdf.

Cole, Thomas. 1836. "Essay on American Scenery." *American Monthly Magazine*, January 1836. Thomas Cole National Historic Site. https://thomascole.org/wp-content/uploads/Essay-on-American-Scenery.pdf.

Frank, Peter W. 1982. "Effects of Winter Feeding on Limpets by Black Oystercatchers, Haematopus bachmani." *Ecology* 63:1352–1362. https://www.jstor.org/stable/1938863?seq=1#page_scan_tab_contents.

Myers, Isabelle Briggs, with Peter B. Myers. 1995 [1980]. *Gifts Differing: Understanding Personality Types*. 2nd ed. Mountain View, CA: Davies-Black/CCP.

Otis Cafe. https://otiscafe.com/.

Tessler, D. F., J. A. Johnson, B. A. Andres, S. Thomas, and R. B. Lanctot. 2007. *Black Oystercatcher* (Haematopus bachmani) *Conservation Action Plan*. International Black Oystercatcher Working Group, US Fish and Wildlife Service. Anchorage: Alaska Department of Fish and Game and Manomet Center for Conservation Sciences. https://www.fws.gov/oregonfwo/Species/Data/BlackOystercatcher/Documents/Black_oystercatcher_conservation_action_plan_FINAL_April07.pdf.

Thoreau, Henry David. 1852. Journal, July 18. https://www.walden.org/wp-content/uploads/2016/02/Journal-4-Chapter-3.pdf.

7. Voices of the Old Forest

Egerton, Frank N. 2011. "History of Ecological Sciences, Part 39: Henry David Thoreau, Ecologist." https://esajournals.onlinelibrary.wiley.com/doi/10.1890/0012-9623-92.3.251.

Franklin, Jerry F., et al. 1968. "Chemical Soil Properties under Coastal Oregon Stands of Alder and Conifers." Proceedings of Northwest Scientific Association Annual Meeting, April 14–18, 1967. https://andrewsforest.oregonstate.edu/sites/default/files/lter/pubs/pdf/pub350.pdf.

Grable, Juliet. 2018. "From Sea to Tree, Scientists Are Tracking Marbled Murrelets with Rising Precision." *Audubon* (Fall). https://www.audubon.org/magazine/fall-2018/from-sea-tree-scientists-are-tracking-marbled.

Lindemann, Berndt, Yoko Ogiwara, and Yuzo Ninomiya. 2002. "The Discovery of Umami." *Chemical Senses* 27 (9): 843–884. https://doi.org/10.1093/chemse/27.9.843.

Spies, Thomas A. 2009. "Science of Old Growth, or a Journey into Wonderland." In *Old Growth in a New World: A Pacific Northwest Icon Reexamined*, edited by T. A. Spies and S. L. Duncan, 31–41. Washington, DC: Island Press.

Spies, T. A., P. A. Stine, R. Gravenmier, J. W. Long, and M. J. Reilly, technical coordinators. 2018. *Synthesis of Science to Inform Land Management within the Northwest Forest Plan Area*. Vol. 1. General Technical Report PNW-GTR-966. Portland, OR: USDA, Forest Service, Pacific Northwest Research Station. https://www.fs.fed.us/pnw/pubs/pnw_gtr966.pdf.

Wolfe, Linnie Marsh, ed. 1979. *John of the Mountains: The Unpublished Journals of John Muir*. Madison: University of Wisconsin Press.

8. So Long, Silverspot

Cascade Head Preserve, Oregon. nd. The Nature Conservancy. https://www.nature.org/en-us/get-involved/how-to-help/places-we-protect/cascade-head/.

Hill, Ryan I., et al. 2018. "Effectiveness of DNA Barcoding in Speyeria Butterflies at Small Geographic Scales." October 15. https://www.mdpi.com/1424-2818/10/4/130.

IDEAS Visualization Team, Oregon State University. nd. "The Oregon Silverspot Butterfly." YouTube video. https://www.youtube.com/watch?v=NI3NVOp5sTg&feature=youtube.

Klag, Graham, Deanna Williams, Cynthia Glick, Margery Price, Ayala Irvin, Brandon Renhard and Preston Nightingale. 2016. "Coastal Prairie Project: Continuing a Vision for a Treasured Landscape." Report prepared for USDA Forest Service, Siuslaw National Forest; Salmon Drift Creek Watershed Council; US Fish and Wildlife Service; Institute for Applied Ecology; and Lincoln County School District.

McCorkle, David V., and Paul C. Hammond. 1988. "Biology of *Speyeria zerene Hippolyta* (Nymphalidae) in a Marine-Modified Environment." *Journal of the Lepidopterists' Society* 42 (3): 184–195. http://images.peabody.yale.edu/lepsoc/jls/1980s/1988/1988-42(3)184-McCorkle.pdf.

McHugh, Anne, Paulette Bierzychudek, Christina Greever, Tessa Marzulla, Richard Van Buskirk, and Greta Binford. 2013. "A Molecular Phylogenetic Analysis of Speyeria and Its Implications for the Management of the Threatened *Speyeria zerene hippolyta*." *Journal of Insect Conservation* 17:1237–1253. https://link.springer.com/article/10.1007%2Fs10841-013-9605-5.

Miller, Mark P., Thomas D. Mullins, and Susan M. Haig. 2016. "Genetic Diversity and Population Structure in the Threatened Oregon Silverspot Butterfly (*Speyeria zerene hippolyta*) in Western Oregon and Northwestern California—Implications for Future Translocations and the Establishment of New Populations." US Geological Survey, Open-File Report 2016–1162. Prepared in cooperation with the US Fish and Wildlife Service. https://pubs.usgs.gov/of/2016/1162/ofr20161162.pdf.

Oregon Silverspot Butterfly. 2019. US Fish and Wildlife Service, Oregon Fish and Wildlife Office. https://www.fws.gov/oregonfwo/articles.cfm?id=149489459. Last modified November.

Pickering, Debbie, and Dan Salzer. 2014. "Adaptive Management on Cascade Head Preserve and CAP for Salmon River Watershed." Open Standards for the Practice of Conservation. http://cmp-openstandards.org/case-study/tnc-cascade-head-preserve-and-salmon-river-watershed/. Accessed May 3, 2020.

Pyle, Robert Michael, and Caitlin C. LaBar. 2018. *Butterflies of the Pacific Northwest.* Portland, OR: Timber Press.

Ripley, James D. 1983. "Description of the Plant Communities and Succession of the Oregon Coast Grasslands." PhD dissertation, Oregon State University. https://ir.library.oregonstate.edu/concern/graduate_thesis_or_dissertations/z029p717g?locale=en.

Sims, Steven R. 2017. "*Speyeria* (Lepidoptera: Nymphalidae) Conservation." Insects 8 (2): 45. https://www.ncbi.nlm.nih.gov/pmc/articles/PMC5492059/. Accessed 8/12/20.

US Fish and Wildlife Service. 2001. "Revised Recovery Plan for the Oregon Silverspot Butterfly (*Speyeria zerene hippolyta*)." https://www.cabi.org/isc/FullTextPDF/2013/20137204849.pdf.

US Fish and Wildlife Service. 2017. "Butterflies Take Wing at Restored Prairie at Oregon Refuge." *Open Spaces,* August, 24. https://www.fws.gov/news/blog/index.cfm/2017/8/24/Butterflies-Take-Wing-at-Restored-Prairie-at-Oregon-Refuge.

US Forest Service. nd. "Oregon Silverspot Butterfly." https://www.fs.usda.gov/detail/siuslaw/learning/nature-science/?cid=fseprd522077.

9. Fish in the Forest

Conservation Biology Institute. nd. "VELMA—A Scalable, Transferable Ecohydrological Model for Watershed Restoration Planning." Webinar. https://consbio.org/products/webinars/velma-ecohydrological-model.

Dietrich, William. 2010. *The Final Forest: Big Trees, Forks, and the Pacific Northwest.* Seattle: University of Washington Press.

Franklin, Jerry F., K. Norman Johnson, and Debora L. Johnson. 2018. *Ecological Forest Management.* Long Grove, IL: Waveland Press.

Frissell, Christopher A. 2017. "Implications of Perry and Jones (2016) Study of Streamflow Depletion Caused by Logging for Water Resources and Forest Management in the Pacific Northwest." Frissell and Raven Hydrobiological and Landscape Sciences. http://oregon-stream-protection-coalition.com/wp-content/uploads/2017/10/MEMO-RE-Implications-of-Perry-and-Jones-2016.pdf.

Google Earth Engine. nd. "A Planetary-Scale Platform for Earth Science Data and Analysis." https://earthengine.google.com/.

Jones, Julia A. 2000. "Hydrologic Processes and Peak Discharge Response to Forest Removal, Regrowth, and Roads in 10 Small Experimental Basins, Western Cascades, Oregon." *Water Resources Research* 36 (9): 2621–2642. https://andrewsforest.oregonstate.edu/publications/2567.

Kennedy Geospatial Lab. nd. "LandTrendr." Oregon State University. http://geotrendr.ceoas.oregonstate.edu/landtrendr/.

Kernville–Gleneden Beach–Lincoln Beach Water District. http://kgblbwater.com/.

Kerr, Andy. 2004. *A Brief Unnatural History of Oregon's Forests*. Portland, OR: Timber Press. http://www.andykerr.net/oregon-wild-the-book.

Lake, Rebecca. 2018. "Diversify Your Portfolio by Investing in Timber." *US News & World Report*, September 7. https://money.usnews.com/investing/real-estate-investments/articles/2018-09-07/diversify-your-portfolio-by-investing-in-timber.

McKane, Bob A., et al. 2014. "Modeling Ecosystem Service Tradeoffs for Alternative Land Use and Climate Scenarios." CES Conference, Washington, DC, December 08–12. EPA Science Inventory. https://cfpub.epa.gov/si/si_public_record_report.cfm?Lab=NHEERL&dirEntryId=299097.

Mid-Coast Water Planning Partnership. 2020. http://midcoastwaterpartners.com/water-on-the-mid-coast/.

Native Fish Society. nd. http://nativefishsociety.org/.

Native Fish Society. 2020. "The Science That Guides Us." https://nativefishsociety.org/science.

NOAA National Marine Fisheries Service. 2016. "2016 5-Year Review: Summary & Evaluation of Southern Oregon/Northern California Coast Coho Salmon." https://repository.library.noaa.gov/view/noaa/17026.

NOAA Fisheries, West Coast Region. nd. "Oregon Coast Coho." https://www.westcoast.fisheries.noaa.gov/protected_species/salmon_steelhead/salmon_and_steelhead_listings/Coho/oregon_coast_Coho.html.

NOAA Fisheries. 2016. "Final ESA Recovery Plan for Oregon Coast Coho Salmon (*Oncorhynchus kisutch*)." https://repository.library.noaa.gov/view/noaa/15986.

Oregon Department of Fish and Wildlife. nd. "Oregon Coast Coho Conservation Plan." https://www.dfw.state.or.us/fish/CRP/coastal_Coho_conservation_plan.asp.

Oregon Department of Fish and Wildlife. nd. "Backgrounder: Instream Water Rights." https://www.dfw.state.or.us/fish/water/docs/BKGWaterRights.pdf.

Oregon Department of Fish and Wildlife. nd. "Water Quality and Quantity Program." https://www.dfw.state.or.us/fish/water/.

Perry, Timothy D., and Julia A. Jones. 2016. "Summer Streamflow Deficits from Regenerating Douglas-Fir Forest in the Pacific Northwest, USA." *Ecohydrology*, Special Issue Paper. https://onlinelibrary.wiley.com/doi/abs/10.1002/eco.1790.

Salmon Drift Creek Watershed Council. http://www.salmondrift.org/#.

10. Beavers in Pixieland

Bouwes, Nicolaas, et al. 2016. "Ecosystem Experiment Reveals Benefits of Natural and Simulated Beaver Dams to a Threatened Population of Steelhead (*Oncorhynchus mykiss*)." *Scientific Reports* 6:28581. https://doi.org/10.1038/srep28581.

Ellingson, Kami S., and Barbara J. Ellis-Sugai. 2014. "Restoring the Salmon River Estuary: Journey and Lessons Learned Along the Way 2006–2014." US Forest Service. https://www.fs.usda.gov/Internet/FSE_DOCUMENTS/fseprd563410.pdf.

Feinstein, Kelly. 2006. "A Brief History of the Beaver Trade." https://humwp.ucsc.edu/cwh/feinstein/A%20brief%20history%20of%20the%20beaver%20trade.html.

Goldfarb, Ben. 2018. Eager: *The Surprising, Secret Life of Beavers and Why They Matter*. White River Junction, VT: Chelsea Green.

Goldfarb, Ben. 2018. "In Oregon, a Peculiar Case for Protecting the Beaver." *High Country News*, February 20. https://www.hcn.org/issues/50.4/wildlife-can-the-beaver-state-learn-to-love-beavers.

Johnston, Carol A. 2015. "Fate of 150 Year Old Beaver Ponds in the Laurentian Great Lakes Region." *Wetlands* 35:1013–1019. https://doi.org/10.1007/s13157-015-0688-5.

Loew, Tracy. 2018. "Feds Suspend Oregon Beaver-Killing Program." *Statesman Journal* 10 (January). https://www.statesmanjournal.com/story/tech/science/environment/2018/01/10/oregon-beaver-killing-program-suspended-federal-officials/1021984001/.

Mills, Enos A. 1909. "The Beaver and His Works." In *Wild Life on the Rockies.* http://www.learning-living.com/2016/07/noble-nature-beaver-and-his-works-by.html.

Mills, Enos A. 1913. *In Beaver World.* https://www.amazon.com/Beaver-World-Enos-Mills/dp/076619387X.

Naiman, Robert J., Carol A. Johnston, and James C. Kelly. 1988. "Alteration of North American Streams by Beaver: The Structure and Dynamics of Streams Are Changing as Beaver Recolonize Their Historic Habitat." *BioScience* 38 (11): 753–762. https://doi.org/10.2307/1310784.

National Park Service. nd. "Enos Mills Naturalist, Father of Rocky Mountain National Park, Homesteader 1870–1922 Estes Park, Colorado." https://www.nps.gov/home/learn/historyculture/upload/MW,pdf,-MillsBio,b.pdf.

Nickelson, Thomas E., et al. 1992. "Seasonal Changes in Habitat Use by Juvenile Coho Salmon (*Oncorhynchus kisutch*) in Oregon Coastal Streams." *Canadian Journal of Fisheries and Aquatic Sciences* 49 (4):783–789. https://martinezbeavers.org/wordpress/wp-content/uploads/2018/01/Nickelson-et-al.-1992-Seasonal-Changes-in-Habitat-Use-by-Juvenile-Coho-Salmon-Oncorhynchus-kisutch-in-Oregon-Coastal-Streams-Can-J-Fish-Aquat-Sci.pdf.

NOAA Fisheries. 2016. "Final ESA Recovery Plan for Oregon Coast Coho Salmon (*Oncorhynchus kisutch*)." https://repository.library.noaa.gov/view/noaa/15986.

Pacific States Marine Fisheries Commission, Habitat Program. nd. "Beaver Benefits and Controlling Impacts." http://habitat.psmfc.org/living-with-beaver/.

UNESCO-MAB. nd. "Shatskyi Biosphere Reserve." http://www.unesco.org/mabdb/br/brdir/directory/biores.asp?code=UKR+04&mode=all.

UNESCO-MAB. 2012. "West Polesie Biosphere Reserve." http://www.unesco.org/new/en/natural-sciences/environment/ecological-sciences/biosphere-reserves/europe-north-america/belaruspolandukraine/west-polesie-transboundary-biosphere-reserve/. Last modified September.

UNESCO-MAB Biosphere Reserves Directory. nd. "Biosphere Reserve Location Map—Ukraine." http://www.unesco.org/mabdb/br/brdir/europe-n/Ukrainemap.htm.

UNESCO Man and the Biosphere Programme. 2016. "Cascade Head Biosphere Reserve Periodic Review." Final/approved version, September 25. US Forest Service, Hebo Ranger District.

USAID/Ukraine. 2017. "Ukraine Biodiversity Analysis." http://www.brucebyersconsulting.com/wp-content/uploads/2017/07/Ukraine-Biodiversity-Analysis-Report-Final-as-Posted-on-DEC.pdf.

Wikipedia. nd. "John Jacob Astor." https://en.wikipedia.org/wiki/John_Jacob_Astor.

Wild Salmon Center. 2020. "Native Beavers Join Oregon Wild Coho Recovery Work." September 28. https://www.wildsalmoncenter.org/2018/09/28/native-beavers-join-oregon-wild-Coho-recovery-work/.

Wilkinson, Charles. 2010. *The People Are Dancing Again: The History of the Siletz Tribe of Western Oregon.* Seattle: University of Washington Press.

11. Where Have All the Seastars Gone?

Byers, Bruce A., and Jeffry B. Mitton. 1981. "Habitat Choice in the Intertidal Snail Tegula funebralis." *Marine Biology* 65:149–154. https://doi.org/10.1007/BF00397079.

Byers, Bruce A. 1983. "Enzyme Polymorphism Associated with Habitat Choice in the Intertidal Snail Tegula funebralis." *Behavior Genetics* 13:65–75. https://doi.org/10.1007/BF01071744.

Byers, Bruce A. 2018. "Ecology, the Humbling Science." *Frontiers in Ecology and the Environment* 16 (3): 139–139. https://doi.org/10.1002/fee.1782.

Donahue, Michelle Z. 2016. "Starfish Baby Boom Surprises Biologists." 2016. *National Geographic,* May 24. https://news.nationalgeographic.com/2016/05/160524-starfish-baby-boom-surprises-biologists/.

Feder, Howard M. 1959. "The Food of the Starfish, Pisaster Ochraceus, along the California Coast." *Ecology* 40 (4): 721–724. https://esajournals.onlinelibrary.wiley.com/doi/abs/10.2307/1929828.

Feldkamp, Lisa. 2016. "Reaching for the (Sea) Stars." *Cool Green Science,* September 13. https://blog.nature.org/science/2016/09/13/reaching-sea-stars-wasting-disease-starfish-citizen-science/.

Harley, Christopher D. G., et al. 2006. "Color Polymorphism and Genetic Structure in the Sea Star Pisaster ochraceus." *Biological Bulletin* 211 (3): 248–262. https://www.journals.uchicago.edu/doi/10.2307/4134547.

Hewson, Ian, et al. 2018. "Investigating the Complex Association between Viral Ecology, Environment, and Northeast Pacific Sea Star Wasting." *Frontiers in Marine Science*, March 7. https://doi.org/10.3389/fmars.2018.00077.

MARINe: Multi-Agency Rocky Intertidal Network. https://marine.ucsc.edu/index.html.

Menge, Bruce A., et al. 2016. "Seastar Wasting Disease in the Keystone Predator *Pisaster ochraceus* in Oregon: Insights into Differential Population Impacts, Recovery, Predation Rate, and Temperature Effects from Long-Term Research." *PLoS One* 11 (5): e0153994. https://doi.org/10.1371/journal.pone.0153994.

Miner, Melissa, et al. 2018. "Large-Scale Impacts of Seastar Wasting Disease (SSWD) on Intertidal Seastars and Implications for Recovery." *PLoS One* 13 (3): e0192870. https://doi.org/10.1371/journal.pone.0192870.

Oregon Department of Fish and Wildlife. nd. "Harvest Restrictions: Cascade Head Marine Reserve and Marine Protected Areas (MPA)." https://oregonmarinereserves.com/content/uploads/2016/02/CascadeHead.pdf.

Pacific Rocky Intertidal Monitoring: Trends and Synthesis. 2019. "Sea Star Wasting Syndrome Map." http://data.piscoweb.org/marine1/seastardisease.html. Last modified November 13.

Paine, Robert T. 1966. "Food Web Complexity and Species Diversity." *American Naturalist* 100 (910): 65–75. https://www.jstor.org/stable/2459379?seq=1.

Paine, Robert T. 1969. "A Note on Trophic Complexity and Community Stability." *American Naturalist* 103 (929): 91–93. https://www.journals.uchicago.edu/doi/pdfplus/10.1086/282586.

Paine, Robert T. 1969. "The *Pisaster-Tegula* Interaction: Prey Patches, Predator Food Preference and Inter-Tidal Community Structure." *Ecology* 50 (6): 950–961. https://esajournals.onlinelibrary.wiley.com/doi/abs/10.2307/1936888.

PISCO: Partnership for Interdisciplinary Studies of Coastal Oceans. http://www.piscoweb.org/.

Raimondi, Peter T., et al. 2007. "Consistent Frequency of Color Morphs in the Seastar *Pisaster ochraceus* (Echinodermata: Asteriidae) across Open-Coast Habitats in the Northeastern Pacific." *Pacific Science* 61 (2): 201–210. https://doi.org/10.2984/1534-6188(2007)61[201:CFOCMI]2.0.CO;2.

Ricketts, Edward F., and Jack Calvin. 1939. *Between Pacific Tides.* Stanford, CA: Stanford University Press.

Rogue Ales and Spirits. 2020. "Wasted Sea Star Purple Pale Ale." https://www.rogue.com/stories/purple-pale-ale-helps-sea-stars. Accessed May 4.

Steinbeck, John. 1951. *Log from the Sea of Cortez: The Narrative Portion of the Book,* Sea of Cortez, by John Steinbeck and E. F. Ricketts, 1941. London: Penguin Books.

Thompson, Francis. 1897. "The Mistress of Vision." First published in *New Poems.* Boston: Copeland and Day. https://www.bartleby.com/336/602.html.

University of Washington, Department of Biology. 2018. "Robert T. Paine's Keystone Species Featured in Short Film." March 9. https://www.biology.washington.edu/news/news/1520632800/robert-t-paines-keystone-species-featured-short-film.

Wikipedia. 2020. "Sea Star Wasting Disease." https://en.wikipedia.org/wiki/Sea_star_wasting_disease. Last modified January 12.

Yong, Ed. 2013. "The Man Whose Dynasty Changed Ecology." *Scientific American,* January 13. https://www.scientificamerican.com/article/the-man-whose-dynasty-changed-ecology/?redirect=1.

12. Whale Haven

Aridjis, Homero. 2002. "El ojo de la ballena" and "The Eye of the Whale." In *Eyes to See Otherwise (*Ojos de otro mirar*): Selected Poems.* Translated by Betty Ferber and George McWhirter. New York: New Directions.

Calambokidis, John, et al. 2012. "Updated Analysis of Abundance and Population Structure of Seasonal Gray Whales in the Pacific Northwest, 1998–2010." https://www.westcoast.fisheries.noaa.gov/publications/protected_species/marine_mammals/cetaceans/gray_whales/studies_under_review/calambokidis_et_al_2012.pdf.

Callahan, Mary. 2019. "As California Gray Whale Deaths Mount, Scientists

Raise Alarm about Ocean Health." *Press Democrat,* May 24. https://www.pressdemocrat.com/news/9632857-181/as-california-gray-whale-deaths.

Darling, Jim, et al. 1998. "Gray Whale (Eschrichtius robustus) Habitat Utilization and Prey Species off Vancouver Island, B.C." *Marine Mammal Science* 14 (4): 692–720. https://doi.org/10.1111/j.1748-7692.1998.tb00757.x.

Diaz, Alexa. 2019. "13th Dead Gray Whale Washes Up in Northern California." *Los Angeles Times,* May 24. https://www.latimes.com/local/lanow/la-me-ln-dead-whale-20190524-story.html.

Guazzo, Regina A., et al. 2017. "Migratory Behavior of Eastern North Pacific Gray Whales Tracked Using a Hydrophone Array." *PLoS One* 12 (10): e0185585. https://doi.org/10.1371/journal.pone.0185585.

King James Bible. nd. Genesis 1:27–28. https://www.kingjamesbibleonline.org/Genesis-1-27/.

Kuyimá Ecoturismo. http://www.kuyima.com/whales/.

Makah Tribal Council. nd. "The Makah Whaling Tradition." https://makah.com/makah-tribal-info/whaling/.

Newell, Carrie. 2016. "Meet the Resident Whales of Depoe Bay." *Portland Monthly,* May 5. https://www.pdxmonthly.com/slideshows/2016/5/5/meet-the-resident-gray-whales-of-depoe-bay.

NOAA Fisheries. nd. "Gray Whale." Species Directory. https://www.fisheries.noaa.gov/species/gray-whale.

NOAA Fisheries. nd. "Gray Whale." Southwest Fisheries Science Center. https://swfsc.noaa.gov/textblock.aspx?ParentMenuId=230&id=1431.

NOAA Fisheries. 2018. "Gray Whale (*Eschrichtius robustus*): Eastern North Pacific Stock." 2018 Report. Marine Mammal Stock Assessment Reports by Species/Stock. https://www.fisheries.noaa.gov/national/marine-mammal-protection/marine-mammal-stock-assessment-reports-species-stock.

NOAA Fisheries. 2019. "Makah Tribal Whale Hunt." 2012 Notice of Intent. https://www.fisheries.noaa.gov/west-coast/makah-tribal-whale-hunt. Last modified June 12.

NOAA Fisheries. 2019. "Makah Tribal Whale Hunt FAQs." https://www.fisheries.noaa.gov/west-coast/makah-tribal-whale-hunt-frequently asked-questions. Last modified August 1.

Oregon State Parks. nd. "Whale Watching." https://oregonstateparks.org/index.cfm?do=thingstodo.dsp_whalewatching.

Pyenson, Nicholas, and David Lindberg. 2011. "What Happened to Gray Whales during the Pleistocene? The Ecological Impact of Sea-Level Change on Benthic Feeding Areas in the North Pacific Ocean." *PLoS One* 6 (7): e21295. https://doi.org/10.1371/journal.pone.0021295.

Renker, Ann M. nd. "The Makah: People of the Sea and Forest." University of Washington Libraries. http://content.lib.washington.edu/aipnw/renker.html.

Russell, Dick. 2001. *Eye of the Whale: Epic Passage from Baja to Siberia*. New York: Island Press.

UNESCO. nd. "Whale Sanctuary of El Vizcaíno." World Heritage List. http://whc.unesco.org/en/list/554.

Western Gray Whale Advisory Committee. 2018. "Report of the Fourth Rangewide Workshop on the Status of North Pacific Gray Whales." *Journal of Cetacean Research and Management* 19 (Suppl.): 521–536. https://www.iucn.org/sites/dev/files/wgwap18_inf6.pdf.

Whale Watching Spoken Here. nd. https://whalespoken.wordpress.com/.

Wikipedia. 2019. "Charles Melville Scammon." https://en.wikipedia.org/wiki/Charles_Melville_Scammon. Last modified October 24.

Wikipedia. 2019. "El Vizcaíno Biosphere Reserve." https://en.wikipedia.org/wiki/El_Vizca%C3%ADno_Biosphere_Reserve. Last modified December 26.

Wikipedia. 2020. "Homero Aridjis." https://en.wikipedia.org/wiki/Homero_Aridjis. Last modified January 16.

13. Salmon in the Net of Indra

Chen, Bi Xia, and Yuei Nakama. 2004. "A Summary of Research History on Chinese Feng-shui and Application of Feng-shui Principles to Environmental Issues." *Kyushu Journal of Forest Research* 57:297–301. http://ffpsc.agr.kyushu-u.ac.jp/jfs-q/kyushu_forest_research/57/57po013.PDF.

Chen, Bixia. 2008. "A Comparative Study on the Feng Shui Village Landscape and Feng Shui Trees in East Asia—A Case Study of Ryukyu and Sakishima Islands." PhD dissertation, Kagoshima University, Japan.

https://www.researchgate.net/publication/38410455_A_Comparative_Study_on_the_Feng_Shui_Village_Landscape_and_Feng_Shui_Trees_in_East_Asia-A_Case_Study_of_Ryukyu_and_Sakishima_Islands-.

Drift Creek Camp. nd. "History." https://driftcreek.org/about-us/history/.

Drift Creek Camp. nd. "About Drift Creek Nature Center." https://driftcreek.org/nature-center/about-dcnc/.

Egerton, Frank N. 2011. "History of Ecological Sciences, Part 39: Henry David Thoreau, Ecologist." https://esajournals.onlinelibrary.wiley.com/doi/10.1890/0012-9623-92.3.251.

Fields, Rick. 1981. *How the Swans Came to the Lake: A Narrative History of Buddhism in America*. Boston: Shambhala.

Marx, Steven. 2013. "Thoreau's Buddhism." A presentation to the White Heron Sangha, June 23. http://www.stevenmarx.net/2013/06/excerpt/#_edn4.

Piez, Wendell. 1992. "The Rain of Law: Thoreau's Translation of the Lotus Sutra." *Tricycle*, Winter. https://tricycle.org/magazine/rain-law/.

Quammen, David. 2018. *The Tangled Tree: A Radical New History of Life*. New York: Simon and Schuster.

Rendell, James R., and I. Tansley, trans. 1912. *The Principia of Emanuel Swedenborg*. Vol. 2, quoted in John G. Burke, *Origins of the Science of Crystals*. Berkeley: University of California Press, 1966.

Scott, David. 2007. "Rewalking Thoreau and Asia: 'Light from the East' for 'A Very Yankee Sort of Oriental.'" *Philosophy East and West* 57 (1): 14–39. https://www.jstor.org/stable/4488074.

Snyder, Gary. 1990. *The Practice of the Wild*. San Francisco: North Point Press.

Thoreau, Henry David. 1962. *The Variorum Walden and the Variorum Civil Disobedience*. Annotated and with Introductions by Walter Harding. New York: Washington Square Press.

Walls, Laura Dassow. 2017. *Henry David Thoreau: A Life*. Chicago: University of Chicago Press.

Worster, Donald. 2008. *A Passion for Nature: The Life of John Muir*. New York: Oxford University Press.

14. Dancing on the Shortest Day

Beckham, Stephen Dow. 1975. *Cascade Head and the Salmon River Estuary: A History of Indian and White Settlement.* Report to the US Forest Service to comply with requirements of the Antiquities and Historical Preservation Acts. Siuslaw National Forest, Hebo Ranger District.

Beckham, Stephen Dow, narrative author. 2018. *Rise of the Collectors.* Grand Ronde, OR: Chachalu Museum and Cultural Center.

Boas, Franz. 1898. "Traditions of the Tillamook Indians." *Journal of American Folklore* 11 (40): 23–38. https://www.jstor.org/stable/533608.

Clayoquot Biosphere Trust. nd. http://clayoquotbiosphere.org/.

Clayoquot Sound UNESCO Biosphere Reserve. nd. http://www.unesco.org/new/en/natural-sciences/environment/ecological-sciences/biosphere-reserves/europe-north-america/canada/clayoquot-sound/.

Confederated Tribes of Grande Ronde. nd. "Chachalu Museum and Cultural Center." https://www.grandronde.org/departments/cultural-resources/chachalu-museum-and-cultural-center/.

Confederated Tribes of Siletz Indians. nd. http://www.ctsi.nsn.us/.

Fixico, Donald. 2019. "Termination and Restoration in Oregon." *Oregon Encyclopedia.* https://oregonencyclopedia.org/articles/termination_and_restoration/#.XRtyhOtKh0w. Last modified July 10.

Lewis, David G. 2009. "Termination of the Confederated Tribes of the Grand Ronde Community of Oregon: Politics, Community, Identity." PhD dissertation, Department of Anthropology, University of Oregon. https://core.ac.uk/download/pdf/36684919.pdf.

Minor, Rick, and Wendy C. Grant. 1996. "Earthquake Induced Subsidence and Burial of Late Holocene Archaeological Sites, Northern Oregon Coast." *American Antiquity* 61 (4): 772–781. https://www.jstor.org/stable/282017?seq=1.

Ortiz, Alfonso. 1969. *The Tewa World: Space, Time Being and Becoming in a Pueblo Society.* Chicago: University of Chicago Press.

Pearson, Clara. 1990. *Nehalem Tillamook Tales.* Told by Clara Pearson; recorded by Elizabeth Derr Jacobs; edited by Melville Jacobs; introduction by Jarold Ramsey. Corvallis: Oregon State University Press.

"Song of the Sky Loom." 1993. In *Songs of the Tewa,* collected and translated by Herbert Joseph Spinden in 1933. 2nd ed. Santa Fe, NM: Sunstone Press. https://www.poetrynook.com/poem/song-sky-loom.

Sweet, Jill D. 1985. *Dances of the Tewa Pueblo Indians*. Santa Fe, NM: School of American Research Press.

Wikipedia. 2019. "Indian Gaming Regulatory Act." https://en.wikipedia.org/wiki/Indian_Gaming_Regulatory_Act. Last modified November 9.

Wilkinson, Charles. 2010. *The People Are Dancing Again: The History of the Siletz Tribe of Western Oregon*. Seattle: University of Washington Press.

Zobel, Donald B. 2002. "Ecosystem Use by Indigenous People in an Oregon Coastal Landscape." *Northwest Science* 76 (4): 304–314. https://research.wsulibs.wsu.edu/xmlui/bitstream/handle/2376/921/v76%20p304%20Zobel.PDF?sequence=1.

15. Out Beyond the Eagle's Kitchen

IWCS. 2015. "Ohlone Basket Weaver—Linda Yamane." YouTube video. June 23. https://www.youtube.com/watch?v=fPBuGZ6sbKw.

KCET. 2020. "The Art of Basket Weaving." *Artbound*. Video. Season 9, Episode 8, April 30. https://www.kcet.org/shows/artbound/episodes/the-art-of-basket-weaving.

Landow, George P. 1988. "Edmund Burke's *On the Sublime*." http://www.victorianweb.org/philosophy/sublime/burke.html. Last modified 1988.

Songs of Love, Luck, Animals, and Magic: Music of the Yurok and Tolowa Indians. 2014. "Basket Song (Yurok)." YouTube video. November 20. https://www.youtube.com/watch?v=7-seI6HnOSc.

West, W. Richard. 2010. "A Basket Is a Song Made Visible." In Wilkinson, *The People Are Dancing Again*, 382. https://books.google.com/books?id=mUDbAwAAQBAJ&pg=PA382&lpg=PA382&dq=The+basket+is+a+song+Mrs.+Mattz&source=bl&ots=vDV4vmukJA&sig=ACfU3U1L_Hpmhh9hc5p-XWRwVpyeSzIllQ&hl=en&sa=X&ved=2ahUKEwjvlvDf9JjjAhVXJ80KHXrRC1MQ6AEwB3oECAkQAQ#v=onepage&q=The%20basket%20is%20a%20song%20Mrs.%20Mattz&f=false.